Bioelectrochemistry I
Biological Redox Reactions

ETTORE MAJORANA INTERNATIONAL SCIENCE SERIES

Series Editor:
Antonino Zichichi
European Physical Society
Geneva, Switzerland

(LIFE SCIENCES)

Bioelectrochemistry I
Biological Redox
Reactions

Edited by

G. Milazzo
University of Rome
Rome, Italy

and

Martin Blank
College of Physicians and Surgeons
Columbia University
New York, New York

Plenum Press • New York and London

6871 - 4762

Library of Congress Cataloging in Publication Data

Main entry under title:

Bioelectrochemistry I.

(Ettore Majorana international science series. Life sciences; v. 11)
"Proceedings of the course on bioelectrochemistry which was the eleventh International School of Biophysics, held November 29–December 5, 1981 in Erice, Sicily, Italy"—Verso t.p.
Bibliography: p.
Includes index.
1. Bioelectrochemistry—Congresses. 2. Oxidation—reduction reaction—Congresses. I. Milazzo, Giulio. II. Blank, Martin, 1933–   III. International School of Biophysics (11th: 1981; Erice, Italy) IV. Series.
QP517.B53B54 1983   574.19'283   83-2443
ISBN 0-306-41340-X

Proceedings of the course on Bioelectrochemistry, which was the Eleventh
International School of Biophysics, held November 29–December 5, 1981,
in Erice, Sicily, Italy

© 1983 Plenum Press, New York
A Division of Plenum Publishing Corporation
233 Spring Street, New York, N.Y. 10013

Printed in the United States of America

PREFACE

This is the first course devoted to bioelectrochemistry held within the framework of the International School of Biophysics.

Although this branch of scientific research is already about two centuries old, as a truly independent one it has been in a stage of lively development since only a few decades ago and this is why a first course at the E. Majorana Center was devoted to it.

Since bioelectrochemistry consists of many sub-fields, it is impossible to include, even superficially, all of them in a short course lasting just a week, and therefore the chapter of redox-reactions was chosen for this first course as being most general in character. But even restricting the course to redox-reactions, only a few subjects could be included and therefore the choice among them was made considering the most general guidelines that could serve as a basis for the further study of individual problems.

In this way we hope to give a sound basis to the study of and to stimulate further interest in this branch of both biological and physical chemistry. This dual interdisciplinary approach is, on the other hand, unavoidable if a more rigorous and logical attack on biological problems in living bodies is to be carried ahead.

CONTENTS

SYMBOLS AND ACRONYMS

For the sake of consistency and to ensure immediate understanding, the symbols of the most commonly occurring quantities and of the most currently used acronyms are collected here. The meaning of some other occasionally used symbols is given in the corresponding text.

Latin alphabet

a	activity	c.s.v.	cyclic scanning voltammetry
a.c.	alternating current	CT	calf thymus
A	absorbance; area	cyt	cytochrome
A	ampere	d	density relative
Acc	acceptor	d	differential (exact)
ADP	adenosine diphosphate	D	diffusion coefficient
Ala	alanine	d.c.	direct current
AMP	adenosine monophosphate	DHAP	dihydroxyacetonphosphate
AMPH	ampholyte	d.m.e.	dropping mercury electrode
AODC	average oxidation degree of a C atom	DNA	deoxyribunucleic acid
AONC	average oxidation numbers of a C atom	DNAase	deoxyribonuclease
		$DNAD^+$	deamino NAD^+
ApA	adenine dinucleotide	DiPGA	1,3 diphosphoglyceric acid
Arg	arginine	DOC	deoxycholate
ATP	adenosine triphosphate	Don	donor
ATPase	adenosine triphosphatase	DPN	s. NAD^+
$[Bchl]_2$	bacteriochlorophyll dimer	DPNH	s. NADH
BLM	bilayer lipid membrane	d.p.p.	differential pulse polarography
BSA	bovine serum albumin	e	electric charge of the electron
c	concentration (general)	e^-	electron
C	capacitance	E	energy general, internal
C	coulomb	$\tilde{E}$	electrochemical internal energy
$^\circ C$	centigrade temperature degree	E	electric field strength
cal	calorie (small)	ECE	electrochemical-chemical-electro-chemical reaction
car	carotenoid		
Chl	chlorophyl	E. coli	escherichia coli
c.d.	current density	EDTA	ethylenediamine tetra acetate
CMP	cytosine mononucleotide	e.p.r.	electron paramagnetic resonance
CoA	coenzyme A	eq	equivalent (chemical)
CpC	cytosine dinucleotide	e.s.r.	electron spin resonance

exp (...)	$e^{(...)}$ (exponential)
eV	electron volt
F	farad
F	formal
F	Faraday's constant (96 500 C)
FAD	flavin adenine dinucleotide
FCCP	carbonylcyanide p-fluoromethoxy-phenylhydrazone
FMN	flavin mononucleotide
(g)	gas
g	gram
G	free enthalpy (Gibbs); conductance
$\tilde{G}$	electrochemical free enthalpy
$G°$	standard free enthalpy
GAP	glyceraldehyde-3-phosphate
GC	guanine-cytosine
h	Planck's constant
h	hour
Hz	hertz
H	enthalpy
$\tilde{H}$	electrochemical enthalpy
HPLC	high performance liquid chromatography
I	current intensity
I_0	exchange current intensity
I_c	capacitive current
I_d, I_{diff}	diffusion current
I_f	faradaic current
I_k	kinetic current
I_{lim}	limiting current
I_p	peak current
i.v.	inverse voltammetry
j	current density
j_0	exchange current density
J	joule
J	flux
k	Boltzmann's constant; rate constant
K	Kelvin (absolute or thermodynamic temperature degree)
K	equilibrium constant
l	liter
l	length
(l)	liquid
Leu	leucine
Lip S_2	lipoic acid, oxidized form
Lip(SH)$_2$	lipoic acid reduced form
ln	natural logarithm (base e)
log	decimal logarithm (base 10)
l.s.v.	linear scanning voltammetry

Lys	lysine
m	meter
m	mass
M	concentration [mole/dm^3 (liter), formerly molarity]
min	minute
mole	quantity of substance
8 MOP	8-methoxy-psoralene
MOPS	morphol*iso*propane sulfonate
n	number of moles; number of electrons
N	concentration [chemical equivalents/dm^3 (liter), formerly normality]
N_A	Avogradro's number
NAD	nicotinamide adenine dinucleotide
NAD$^+$	nicotinamide adenine dinucleotide oxidized form
NADH	nicotinamide adenine dinucleotide reduced form
NADP$^+$	triphosphopyridine nucleotide
NADPH	nicotinamide adenine dinucleotide 3' phosphate
n.h.e.	normal hydrogen electrode
NHI	non-heme iron
n.m.r.	nuclear magnetic resonance
n.p.p.	normal pulse polarography
ODC	oxidation degree of a C atom
Ox	oxidant
p	pressure, probability
P	permeability
P$_{870}$	reaction center equivalent to [Bchl]$_2$
PES	phenazine ethosulfate
PGA	3-phosphoglyceric acid
Pi	inorganic phosphate
PMS	phenazine methosulfate
PUVA	psoralene therapy by UV irradiation
Q	quantity of electricity
r	radius
R	gas constant; resistance
r.d.e.	rotating disk electrode
Red	reductant
RNA	ribonucleic acid
RNAase	ribonuclease
s	second
S 13	2',5-dichloro-3-t-butyl-4'-nitrosalicyl-anilide
s.c.e.	saturated calomel electrode
SDH	succinate dehydrogenase
SERS	surface enhanced Raman scattering

(sol)	solution		
t	time, transference number		
T	temperature		
Tris	tris (hydroxymethyl)aminomethane		
TMPD	N,N,N',N',tetramethylparaphenylene diamine		
TOD	total oxidation degree		
TONC	total oxidation number of all C atoms		
TPP	thiamine pyrophosphate		
u	electric mobility		
U	potential of a galvanic cell, or of an electrode		
U°	standard electric potential		
U.V.	ultra violet		
V	volt		
V	volume		
w	weight; work		
W	watt		
z	charge number of an ion		
Z	impedance		

Greek alphabet

α	degree of dissociation
Γ	surface concentration
Δ	difference
ϵ	dielectric constant; molar absorption coefficient; extinction coefficient
λ	wavelength
μ	chemical potential
$\tilde{\mu}$	electrochemical potential
ν	frequency
Σ	sum
σ	surface charge density
τ	relaxation time; drop time
ϕ_ν	quantum yield
Φ	internal (Galvani) electric potential
ψ	external (Volta) electric potential; potential general; electrostatic potential
Ω	ohm

OPENING ADDRESS

ANTONIO BORSELLINO

Istituto di Scienze Fisiche dell'Università
Viale Benedetto XV – Genova 16132, Italy

It is my pleasure and my duty to welcome you here in ERICE. First let me say that, referring to the weather conditions, I hope there will be no further escalation in the way you have been received. If you remember yesterday's metereological events, I think you had already the worst, the snail, a quite rare event in Erice. I welcome you on behalf of the Ettore Majorana Centre for Scientific Culture and of its School, the International School of Biophysics. I wish your work here to be pleasant and fruitful.

I will make a few comments about the present 11th Course of the School, the first in Bioelectrochemistry. When Professor Milazzo proposed the idea to have a Course on this subject, I must admit that I knew very little of the *official* existence of the field of Bio-electrochemistry. I was surprised also to realize how many people there were in a field not very far from mine, of which I knew very little. Correctly I should make some of the comments at the end of the Course, not at the beginning. But I will take the same this opportunity because, I am sorry, I will not be able to stay until the end. About my initial surprise: whatever doubts I could have, they were solved completely by what Professor Milazzo has done, in collaboration with Professor Blank, the other Director of the Course. Professor Milazzo certainly has shown quite a tenacious nature in solving all the difficulties he was confronted with in starting this initiative. First of all he certainly was able to obtain the support of different sources and just now everything seems to be working well. So I myself take the opportunity to thank the Institutions that have supported his efforts. A special mention must go to Council of Europe, that has sent here a representative. The given support confirms how little I knew of Electro-chemistry, or if you wish, of Bio-electro-chemistry. Well, I remember when Biophysics, at least in Italy, started, there were many existential problems. What is Biophysics? So we always had to begin with some philosophical discussion. But later many of these discussions disappeared because Biophysics in the meantime was already in existence. So I am not involving myself in asking what is Bio-electro-chemistry. Someone can give a very simple answer. For example Professor Zichichi, who is the Director of the Majorana Centre, has expressed the view that the all of chemistry and of bio-chemistry, and if you wish, of neurophysiology, everything is simply a special case of electrodynamics, in fact of applied electrodynamics, because all the forces involved are electric. We have nothing to do with the deeper structure of matter. We agree: the living systems are certainly electric, they enter fully in this classification. So we can say: bio-electro-chemistry is just life.

Leaving apart a generality that gives us little comfort, I want to comment on the reasons why there is some separation between biophysics and bio-electro-chemistry. I have the impression that it can be explained in a good measure by the contingency of the academic origin of people. For example, in this course most people are in some way attached to Chemistry, they belong to a Chemistry Department. In Biophysics instead many people are coming from a Physics Department. So we have such difference in the academic formation and the place where people are working. But I don't think there is a deep difference in the nature of problems or on the methods they are using. Therefore this Course, and similar other Courses, will be useful also if they succeed in bringing together people from different academic origin or places of work, putting them under the stress of confronting themselves with people from different origin. Interdisciplinary courses are quite important in this sense and we understand the reason of the support of the European Council, that is very much interested in all interdisciplinary initiatives. We can be confirmed in this view by confronting the Course on Photoreceptors held here last July and the Course that is starting this morning. Lecturers and participants of the previous Course were coming mostly from physics or from department of animal physiology. In the present Course most people are coming from chemistry or plant physiology. The two categories are differentiated by their motivation. For example: photoreceptors are machineries that have been developed to detect photons. They spend the energy already accumulated in the cell to amplify the initial signals elicited by the photons. Instead the main interest of bioelectrochemists is to look to phenomena of the opposite type, those in which photons are used to store energy and this explains the strong connection with the field of bioenergetics. So we have important differences between the two categories. Perhaps we can think that we do not have a basic difference in the nature of the phenomena, but only in the order of magnitude, in the scale on which the two machines are working. From this point of view it will be very important to make the effort to bring together the two clans of peoples and see how much the results in one sector can help people working in the other. In the Course held in July, on the photoreceptors, we had few bio-chemists, Professor Kuhn from Jülich and Professor Bownds, from Wisconsin, and they showed the involvement of phosphodiesterase and of c-GMP in the photoresponses. But very little is known really about what these molecular species are really doing. We had just a starting and it is something that gives a very unpleasant feeling to the biophysicists, to realize that possibly a lot of biochemistry will penetrate the field. I am afraid that the fact must be accepted if we want to have important progress. Many lectures scheduled in the present Course could have equally well been given in the other Course. So I see that there is some partial parallelism between the two Courses and I think that it would be very appropriate to think to organize in the future courses or meetings in the two different directions and make them converging in some way. There is a very famous italian idea of convergent parallels. In some way we must use a non-Euclidean strategy to look for important results, apart from the logic.

I conclude now, wishing the best for your activity here in Erice. If in the future there will be an effort in one or the other sector, I will do what I can to push towards some partial convergence of the two fields, in some aspects separated, but for problems and methods and interests probably more unified than they appear now.

BIOELECTROCHEMISTRY AND BIOENERGETICS
AN INTERDISCIPLINARY SURVEY

GIULIO MILAZZO

Istituto di Chimica della Facoltà di Ingegneria dell'Università di Roma
Via del Castro Laurenziano 7 — 00161 Roma, Italy

An appropriate first approach to an understanding of what we mean by *bioelectrochemistry* lies in its etymology: two prefixes, *bio* and *electro*, indicate the interdisciplinarity of this branch of scientific research, which investigates biological events by using electrochemical techniques, principles and theory.

But this definition is perhaps still too broad, and we must try to restrict it by considering the complexity of biological events, which can be roughly divided in two major groups. The first collects biological events essentially from the point of view of morphology and life functions, but without neglecting environment, which more or less conditions many biological phenomena. Examples of this group are cell subdivision and proliferation, morphology of organs, organization and differentiation of tissues and organs etc. The second group includes the essential life processes from a physicochemical point of view, *i.e.* physicochemical processes essential to life, developed on a molecular basis. Examples of this group are the respiratory chain (*i.e.* combustion — using appropriately transferred atmospheric O_2 — of biological materials in living bodies through a series of redox reactions), heredity (*i.e.* transmission of given properties by means of appropriate combinations and sequences of relatively low molecular species contained in the genes), chemically induced cell proliferation, membrane phenomena (for example regulation of inflow and outflow of charged particles — ions — into and from living cells), anabolic and catabolic processes accumulating and consuming, respectively, the energy needed for life processes (photosynthesis, phosphorylation, aminoacid and protein chemistry, etc.), transmission of information (electrical pulses) by means of, through and to organized structures (nerves, nerve-systems, muscles etc.). All these phenomena are essentially of an electrochemical nature and provide a better basis for the definition of *bioelectrochemistry*.

Bioelectrochemistry can therefore be defined as the branch of scientific research which investigates electrical phenomena resulting from processes involving electrically charged particles (possibly also uncharged particles) belonging to biological systems. It therefore enables a large number of biological events and phenomena to be studied, including the energetics of living bodies, by using strict electrochemical techniques and theories, including solid state physics (semiconductors). The obvious conclusion is that electrochemists, biologists, biophysicists and electrophysiologists must work in very close cooperation, because otherwise any further development would just remain a beautiful dream, but just a dream.

Perhaps unconsciously, but certainly in this spirit scientists have been working in bioelectrochemistry for at least two centuries, and it seems appropriate to describe by means of some examples how scientists of widely differing education have greatly contributed to the development of bioelectrochemistry, thus demonstrating its interdisciplinarity and the need for cooperation.

First of all we have the physician LUIGI GALVANI (born 1737), who with his famous frog experiment (1792) can be considered the father of bioelectrochemistry. At about the same time the physicist ALESSANDRO VOLTA (born 1745) turned his attention to electrical phenomena in living beings, considering nerves to be electric conductors. A few years later JOHANN WILHELM RITTER (born 1776) carried out investigations in electrophysiology and in the field of redox reactions. More recently another physician, LEONOR MICHAELIS (born 1875), and three chemists, DAVID KEILIN (born 1887), RENÉ BERNARD WURMSER (born 1890) and ALBERT SZENT GYÖRGY (born 1893), made extremely important contributions. The first by introducing for the first time the concept of quantitative redox reactions in the investigation of biochemical events involving ionic species; the second with the first models of respiratory processes, involving redox chains; the third, following in MICHAELIS's footsteps, by introducing potentiometric techniques, in the investigation of biological redox reactions; and the fourth by using the theoretical concepts of semiconductors and solid state physics to study the electric conductance of proteins and other biological macromolecules, and hence of organized biological structures. Even in our time ILIA PRIGOGINE (born 1917) has produced very advanced synthesis of theoretical work carried out in the past and introduced some very original sophisticated concepts (local equilibria, dissipative structures, etc.). With the help of a large mathematical framework he was able to prove a substantial measure of agreement between theoretical results and experimental findings for events occurring in states far from thermodynamic equilibrium (among them the majority of biological processes), thus overcoming the apparently unresolved contradiction between the second principle of thermodynamics and the energetics of living bodies. For this he obtained the NOBEL Prize in 1977.

The following few examples will illustrate how varied are the fields involving electrochemistry in biological research.

Chemical reactions in living bodies are of the redox type in which an oxidizable biological system is oxidized for a definite purpose by another biological system present at the same time in the organism. One of the most characteristic examples is the combustion of glucose derivatives in mammals by oxygen carried by hemoglobin. This complete reaction to CO_2 and H_2O actually occurs in a number of individual steps through a series of intermediates. Also other biological substances possibly containing S, P, N etc. atoms can be *burnt* in a similar way, producing corresponding derivatives. This combustion reaction is precisely what supplies the energy needed for the preservation of life. Redox systems of this kind are very difficult to study in their entirety. It is easier to follow them by using electrochemistry after splitting the overall reaction into simpler one-step partial reactions, and this is also probably what really happens. Practically all these simpler one-step redox systems can be studied electrochemically by using the laws of thermodynamics and kinetics. By using the values of redox electric tension, (potential) thermodynamics provides criteria for judging whether and under what conditions a given redox reaction may occur in a living body. For example the oxidation of glucose by oxygen (carried by hemoglobin) can be electrochemically investigated by means of the following partial system (1) and (2)

$$C_6H_{12}O_6 + H_2O \longrightarrow C_6H_{12}O_7 + 2 H^+ + 2 e^- \tag{1}$$

$$2 H^+ + 2 e^- + 1/2 O_2 \longrightarrow H_2O \tag{2}$$

$$\overline{C_6H_{12}O_6 + 1/2 O_2 \longrightarrow C_6H_{12}O_7} \tag{3}$$

the sum of which [equation (3)] is the biological event we wish to investigate.

Reactions (1) and (2) can be easily investigated potentiometrically, thus providing a realiable figure for the standard free enthalpy (*) or reaction (3). Similar splittings for two other biological systems will provide the corresponding figure for the free enthalpy of direct oxidation of the oxidizable system by another reductible system both present at the same time in the given ambient. The comparison of the figures thus obtained will provide a certain indication whether under the given condition the first system will be oxidized or not by the second system.

The kinetic study, on the other hand, which provides information on the corresponding activation energies, and the further introduction of other considerations

(*)   *Free enthalpy* (the logically more correct expression for this quantity) corresponds to what is also otherwise called GIBBS *free energy*.

(for example the analogous of the FRANCK-CONDON (*) principle) enables the time evolution of the redox system investigated to be followed. By using an appropriate conductor with electronic conduction (Pt, Au, W, C etc.) and an appropriate reference electrode, each one of the partial systems, for example system (1) and (2), becomes galvanic cells which enable the values of the standard electric tension and of the rest equilibrium electric tension needed for the computation of the reaction free enthalpy (thermodynamic quantities) to be measured. By allowing the galvanic element to work under the cathodic and anodic load of the redox electrode under investigation, the measured overtension will provide the information needed for computing the activation energy and the reaction rate (kinetic quantity). Obviously we do not find galvanic elements with metallic terminals in living beings, but on the basis of the information obtained under given conditions concerning the free enthalpy (from the electric tension) of the partial systems it is possible to ascertain immediately which of the partial systems — constituting a global redox system — is the oxidizing one and which the oxidized one. On the basis of the activation energies it will also be possible to evaluate, even if approximately, whether the reaction will be fast or slow.

This is a simplified presentation of the biological importance of redox reactions and their possibility of being investigated electrochemically. Other considerations concerning biological redox reactions (presence of reactants in a not completely free state, energetic contribution from superficial adsorption phenomena, use of activities instead of concentrations, classification of biochemical complex reactions in few basic simpler reactions etc.) should be introduced to obtain the best possible results.

The application of concepts and theories of solid state physics and semiconductors to the solution of biological problems is relatively recent (SZENT GYÖRGY 1941). Since then a remarkable amount of research has been carried out which lead to the conclusion that electrical conduction in biological macromolecules can be interpreted on the basis of such concepts. Many such macromolecules (proteins,

---

(*)   The FRANCK-CONDON principle states that if a change in the electronic state of a molecule occurs, for example through absorption or emission of a photon of appropriate energy, the jump of the electron from the initial to the final orbital occurs in a practically not-measurably short time, so that the much heavier (in comparison to the electron) nuclei of the atom constituents of the molecule have no time to move into the equilibrium positions of the new electronic state and therefore remain in the same relative position in which they were before the electronic transition until a new electronic or vibrational transition occurs. In analogy to this principle it is reasonable to think that redox processes (gain or loss of one or more electrons) of polyatomic ions or molecules having as final products another polyatomic ion or molecule whose atomic topology differs from the initial one are in general more difficult and can even become kinetically impossible (from an electrodic point of view).

phthalocyanines, hemoglobin, nucleic acids etc.) show semiconductor properties in the solid state, even if not completely dry. For example in many proteins there are three energy bands lying close to each other (experimental and theoretical findings), of which two are occupied by electrons while the third one, barely 3 eV higher, is empty and can therefore be used as semiconductor conduction band. Excellent results were obtained on this basis for a number of biological events: mitochondrial activity and respiratory chain, behavior of chlorophyll and photosynthesis, mechanism of conduction through organized structures like nerves or muscles etc.

An extremely broad field is that of membranes, *i.e.* of organized structures used by nature to maintain the internal conditions essential for their functions in every cell, *i.e.* for life.

In living bodies the membrane acts as a barrier separating two regions and must be able to open and to close for the transport of matter. One of the mechanisms of transport of matter through membranes operates by way of selective permeability. In this case if the transported particles are electrically charged a number of electric and electrochemical phenomena occur which are consequential and in turn regulate the process. The regions separated by a membrane are usually isoosmotic and therefore the transport of matter must be governed by forces other than osmotic ones; this indicates that the mechanism must be an electrochemical one, and in fact electric potential differences can be measured across biological membranes separating two isoosmotic regions containing ionic species selectively transported.

A number of theoretical biologists are studying these events, starting from basic theories of electrolytic conduction and using the thermodynamics of irreversible processes.

Taking into consideration the non-linearity of gradients of electrical, chemical, and electrochemical potentials (gradients intended as generalized forces operating the investigated transport of matter), the reciprocal interactions *i.e.* the coupling of simultaneously present quantities and occurring phenomena, and using appropriate non-linear differential equations, the existence of negative conductances can be shown, *i.e.* migration of some electrically charged species in a direction opposite to the one apparently predictable by the gradients of electric and electrochemical potential across the membrane. This provides a physical, non-hypothetical meaning to the concept of ionic pumps.

This brillant result in the field of the electric conductance of membranes together with the results obtained by PRIGOGINE and his school in the field of energetics are an indication of the degree to which modern theoretical considerations, apparently very far from biology, can help in the interpretation of biological events.

A further alteration in transmembrane potentials can also have macroscopical consequences, such as those appearing in anesthesia and electro-anesthesia: it is highly probable that in such cases the drugs used modify the electric conductance of cellular membranes for charged species, *i.e.* for ions and possibly for some other kinds of larger molecules (which could also exert a pharmacological action) carrying ions, thus producing some specific material unbalance and the observed anaesthetic phenomena. A typical, and perhaps one of the best-known examples of a relatively large molecule capable of permeating through a membrane because it is bound or complexed with an ion is valinomycin complexing a K^+ ion in the cavity formed by its *bracelet* structure and thus carrying globally a unitary positive charge per molecule.

Membrane concepts also clarify the conducting properties of complex biological structures, as in the case of the transmission of so-called *action potentials,* *i.e.* potential differences generated by a certain excitation, which are transmitted from the peripheral nervous sensors to the central nervous system and then travel back with certain codified information and eventually with an operative order through and to organized complex structures (nerves and muscles). However, the membrane concepts are not sufficient for the interpretation of electric conductance through these complex organized structures. On the basis of the ascertained speed of transmission of the electrical impulses, which is in the order of magnitude of tens of meters per second, it must in fact be concluded that the global conductance through organized structures very probably involves different mechanisms of the ionic, electronic and semiconductor type.

Electrochemical techniques and especially voltammetric ones (including the many variants derived from original simple voltammetry) are very useful in conformational analysis. The denaturation of proteins is but one example of the importance of this kind of investigation. It is known that the denaturation of proteins and nucleic acids is provoked primarily by a change in the shape of the helices constituting these macromolecules, and may involve the breaking of disulfide bonds and/or other kinds of bonds, such as hydrogen bonds. This change of shape, *i.e.* of conformation, can occur when these macromolecules come into contact with charged surfaces, and this in turn can be investigated precisely by voltammetric techniques.

Another very important field of bioelectrochemistry is photosynthesis, *i.e.* the synthesis of biologically important substances using light energy to carry the general reaction

$$A + H_2D + h\nu \longrightarrow H_2A + D$$

where A stands for an acceptor of electrons and D for a donor of electrons. This is of course a redox reaction. A very common example of such redox reactions is the photosynthesis of carbohydrates, carried out in green plants in their chloroplasts by several chlorophyll pigments. Photosynthesis thus represents the ultimate source of energy required for the life processes of practically all living species, either directly (green plants, algae and several bacteria) or indirectly (organisms nourished with products synthesized through photosynthesis by other living species). This example will be developed in more detail during the course.

The last example (in order of mention but not of importance) is the interaction between living cells and electric currents applied directly or generated inductively by applying a changing electromagnetic field. Our knowledge of this area is still extremely limited, but some facts have been proved beyond dispute and led to very important clinical and therapeutical applications. In many cases of bone fracture considered untreatable by orthopaedic means, the use of very small pulsing currents (some μA cm^{-2}) or of alternating electromagnetic fields of appropriate form and dimension (which in turn generate the desired alternating currents) resulted in a very high percentage of complete recoveries (over 80 %).(*)

The table below contains examples of some extreme cases of congenital or traumatically acquired pseudoarthroses to illustrate how effective therapeutic treatment based on bioelectrochemical events can be, as mentioned later on. The final result in all cases was functional union with progressive medularization.

Table 1. — Some extreme cases of pseudoarthroses treated with electromagnetic pulsing fields.

Patients	Age years	Site	Type	Operations	Infected	Amput. recomm.	Treatment months
1	12	Tibia	Congen.	12	—	Yes	3
2	19	Radius	Congen.	3	—	No	4
3	3	Tibia	Congen.	4	—	Yes	2
4	47	Fibula	Acquir.	—	No	No	4
5	38	Shoulder	Acquir.	3	Yes	No	4
6	34	Tibia	Acquir.	4	Yes	Yes	7
7	62	Femur	Acquir.	1	Yes	Yes	6

(*)   This figure is based on statistics published in 1977. It is probable that better results have since been achieved without adverse collateral effects.

It was initially thought that the use of such currents had a stimulating and organizing effect. Now, after a certain number of experiences *in vivo* and *in vitro*, one can assume that both actions exist. Furthermore, an abundant production of free Ca^{2+} ions by bone cells of chicken embryos was experimentally established under the above-mentioned electric stimulation. Only by using the same energy and the same time-distribution were the same results obtained *in vivo* and *in vitro*. On the other hand the absence of therapeutic results was found to parallel the absence of production of free Ca^{2+} ions. The production of free Ca^{2+} ions is related to other biological phenomena like mitosis, transcription of genetic codes, calcification etc. These observations lead to the conclusion that the repair of damaged bone-tissue under electric stimulation is due to electrochemical action as well.

Although this brief survey is inevitably incomplete, it clearly sheds light on the vast area of biological phenomena which can and should be studied with electrochemical techniques if the efforts to increase our knowledge of biological phenomena are to be based on increasingly solid foundations.

Bioelectrochemistry is at present in its initial stage, mainly because of the difficulty of carrying out straightforward experiments on something as complex as a living system. But it is undoubtedly a field which is bound to lead to the development of electrochemical thinking and experimentation.

This is why the first international conference on bioelectrochemistry was organized in Rome in 1971, and it was so successful that it was repeated at Pont à Mousson in 1973, at Jülich in 1975, at Woods Hole (Mass.) in 1977, at Weimar in 1979, in Israel in 1981. The next one will be held at Stuttgart in 1983.

A journal and a number of publications especially devoted to bioelectrochemistry have been founded; the journal *BIOELECTROCHEMISTRY and BIOENERGETICS* is in its ninth year and is now being published as a section of the *JOURNAL OF ELECTROANALYTICAL CHEMISTRY*(*). Four volumes in the series *TOPICS IN BIOELECTROCHEMISTRY AND BIOENERGETICS*(**) have already been published and the fifth is in print.

Additionally, the international *Bio-Electrochemical Society* (BES) has been founded to provide scientists interested in this new branch of scientific research —

(*)   Elsevier Scientific Publishing Company, Amsterdam, Oxford, New York.
(**)   John Wiley and Sons Publisher, Chichester, New York, Brisbane, Toronto.

which promises such great benefits for mankind — an active forum where they can meet and discuss their problems. (*)

The first course on Bioelectrochemistry in the frame of the International School of Biophysics within the E. MAJORANA International Center at Erice, must be welcomed as a move designed to bridge the gap between scientists with a biological or a physicochemical background respectively, to start forming a new generation of scientists with a real interdisciplinary background and stimulate interest and active research work in general in this field. It would be a great achievement if this course were to be followed by other courses in other areas of bioelectrochemistry.

No bibliographic references appear in this lecture, in view of the fact that a long list would be required for each of the fields covered. For some of the subthemes, readers who are interested will find suitable references at the end of the other lectures of this course, while extensive bibliographic details concerning other subjects will be found at the end of each separate article appearing in the various volumes of *Topics in Bioelectrochemistry and Bioenergetics*.

The following monographic chapters have been already published in the *Topics* series. They can be advantageously read as a good introduction in each one of the themes.

Volume 1

M.J. ALLEN, Aspects of electrochemical investigation of biological phenomena.
H. BERG, Polarographic possibilities in protein and nucleic acid research.
R. BUVET, Energetic structure of metabolism.
P.J. ELVING, Nitrogen heterocyclic compounds: electrochemical information concerning energetics, dynamics and mechanisms.
V.S. VAIDHYANATHAN, Philosophy and phenomenology of ion transport and chemical reactions in membrane systems.

Volume 2

YU.A. CHIZMADZHEV and V.F. PASTUSHENKO, Mechanisms of membrane excitability.
M. CIGNITTI, Electrokinetic phenomena in biology.
H.G.L. COSTER and J.R. SMITH, A potential-controlled transient gating mechanism in fixed membrane modules.

(*)   For information and/or application forms write to the Secretary General Prof. Dr. H.W. NÜRNBERG, Institut für Chemie IV Angewandte Physikalische Chemie, Kernforschungsanlage D 517, Jülich, G.F.R., or to the President Prof. Dr. G. MILAZZO, Piazza G. Verdi 9, 00198 Rome, Italy.

S. HJERTÈN, Analytical electrophoresis.

P.G. KOSTYUK, Electrical events during active transport of ions through biological membranes.

CR. SIMIONESCU, SV. DUMITRESCU and V. PERCE, Semiconducting biopolymers and their part in biochemical phenomena.

Volume 3

P.J. ELVING and B.B. GRAVES, Activation energy: nature, determination, significance.

J. KORYTA and M. GROSS, Electrochemistry of complexes of macrocyclic ion carriers and their models.

P. GAZZOTTI, Mitochondria: a general survey.

F. von STURM, Implantable electrodes.

N. RAMASAMY, T.R. LUCAS, B. STANCZEWSKI and P.N. SAWYER, Electrochemical phenomena in vascular homeostasis

M. DELMOTTE and J. CHANU, Non-equilibrium thermodynamics and membrane potential measurement in biology.

Volume 4

V.S. VAIDHYANATHAN, Physico-chemical aspects of homeostasis.

SHU CHIEN, Electrochemical interactions and energy balance in red blood cell aggregation.

E. NEUMANN, Principles of electric field effects in chemical and biological systems.

I.R. MILLER, Structural and energetic aspects of charge transport in lipid layers and in biological membranes.

B.A. FEINBERG and M.D. RYAN, Electron transferring proteins: electrochemical approaches and kinetic-ionic strength effects.

G. DICKEL, Electro-mechanical equilibrium in membranes.

Volume 5

J. KUTA and E. PALECEK, Modern polarographic (voltammetric) techniques in biochemistry and molecular biology. Part 1: Theory and applications in the analysis of low-molecular weight substances.

E. PALECEK, Modern polarographic (voltammetric) techniques in biochemistry and molecular biology. Part 2: Analysis of macromolecules.

F.A. SIDDIQI and H. TI TIEN, Electrochemistry of bilayer lipid membranes.

M.R. TARASEVICH and V.A. BOGDANOVSKAYA, Bioelectrocatalysis. Enzymes as catalysts of electrochemical reactions.

E. SCHOFFENIELS and D.-G. MARGINEANU, Phenomenological and molecular aspects of bioelectrogenesis.

GENERAL CRITERIA FOR THE FULFILMENT OF REDOX REACTIONS

RENÉ BUVET

Laboratoire d'Energétique Electrochimique et Biochimique
Université Paris Val de Marne, Avenue Général De Gaulle
F-94010, Créteil Cédex, France

Contents

1. Introduction

The physicochemical explanation of any kind of reaction occurring in aqueous media and the prevision of the conditions determining this occurrence must always be developed in three successive steps:

— The mass and charge balances of the specific kind of reactions must be clearly settled.
— The energy balances of these reactions must be calculated: first of all for the principle of this evaluation, and then for the numerical values involved in any particular reaction of the kind in question. From the free enthalpy balances associated with the transformation of reagents to products respectively taken in their standard states, equilibrium constants and equilibrium states can be foreseen.
— Finally, kinetic restrictions forbidding or slowing down the accomplishment of the reactions when they are stoichiometrically and energetically possible must be examined, and explained by considering molecular mechanisms.

Concerning these items, redox reactions present some peculiarities, when performed electrochemically by electron transfer between dissolved or adsorbed reagents (or water itself) and electrodes, as well as when the electron exchange occurs chemically between two reagents. We will review these peculiarities, stressing the case of biochemically relevant redox reactions.

2. Mass and charge balances of biochemical redox reactions

2.1. Definitions of redox reactions and redox couples

Oxydo-reduction, or redox, reactions are usually defined as chemical transformations which involve an exchange of electrons. This implies that they can be globally described by stoichiometrically summing up:

— the capture of one or several electrons by a reagent called the oxidant of the electron-accepting redox couple (1):

$$a_1 \, Ox_1 + n_1 \, e^- + p_1 \, H^+ \rightleftharpoons b_1 \, Red_1 + q_1 \, H_2O \qquad (1)$$

—   the supply of one or several electrons by another reagent called the reductant of the donor redox couple (2):

$$b_2 \text{ Red}_2 + q_2 \text{ H}_2\text{O} \rightleftharpoons a_2 \text{ Ox}_2 + n_2 e^- + p_2 \text{ H}^+ \qquad (2)$$

Some redox couples involve other compounds, such as NH_3 in aminating reductions of α-ketoacids or aldehydes, and HScoA in thioesterifying oxidations of aldehydes. But such cases are better treated by complementary modifications of the theory involving additions of non-redox processes to redox couples defined as proposed above. In fact, we have to distinguish, particularly when discussing the kinetics of redox reactions, different kinds of stoichiometry of redox couples as to whether they involve:

—   either simply one or several electrons, as:

$$\text{Ox} + n\,e^- \rightleftharpoons \text{Red}$$

such as in:

$$\text{Fe}^{3+} + e^- \rightleftharpoons \text{Fe}^{2+}$$

—   or a proton-exchange in addition to the electron exchange, as:

$$\text{Ox} + n\,e^- + p\text{H}^+ \rightleftharpoons \text{Red}$$

where p is always positive, such as in:

$$\text{Quinone} + 2\,e^- + 2\,\text{H}^+ \rightleftharpoons \text{Hydroquinone}$$

$$\text{NAD}^+ + 2\,e^- + \text{H}^+ \rightleftharpoons \text{NADH}$$

—   or a modification of the covalent vicinity (topology) of relatively heavy atoms (*i.e.* atoms other than H, such as C, O, N, S, *etc.*) which corresponds to the most general stoichiometry mentioned above, *e.g.*:

$$R_1 R_2 R_3 \equiv \text{COH} + 2\,e^- + 2\,\text{H}^+ \rightleftharpoons R_1 R_2 R_3 \equiv \text{CH} + \text{H}_2\text{O}$$

$$R\text{--CO}_2^- + 2\,e^- + 3\,\text{H}^+ \rightleftharpoons R\text{--CHO} + \text{H}_2\text{O}$$

$$R\text{--CO}_2^- + \text{CH}_3\text{--COScoA} + 2\,e^- + 3\,\text{H}^+ \rightleftharpoons R\text{--CHOH--CH}_2\text{--COScoA} + \text{H}_2\text{O}$$

$$\text{RSSR} + 2\,e^- + 2\,\text{H}^+ \rightleftharpoons 2\,\text{RSH}$$

In aqueous media, the solvated (hydrated) electron has extremely strong reducing properties [$U^\circ = -2.7$ V (s.h.e.)] and therefore cannot exist as such at noticeable concentrations because it should produce gaseous hydrogen by reduction of H_2O. Consequently, the balance of electrons given by Red_2 must equal the balance of electrons captured by Ox_1 and the redox reaction between both has to be written:

$$n_2 a_1 \ Ox_1 + n_1 b_2 \ Red_2 + (n_2 p_1 - n_1 p_2) \ H^+ \rightleftarrows n_2 b_1 \ Red_1 + n_1 a_2 \ Ox_2 +$$

$$+ (n_2 q_1 - n_1 q_2) \ H_2O$$

2.2. *Charge balance of carbonaceous compounds transformations*

Writing the charge balance of global redox couples involving carbonaceous substrates is generally made without any method, but it can be systematized in two different ways.

2.2.1. Oxidation degrees

First, for this purpose, the concept of the oxidation degree of carbon atoms and organic molecules may be introduced. This concept is not as currently used now as it should be in chemistry to define the oxidoreduction state of carbonaceous derivatives, but, in fact, its use is particularly simple as it concerns biochemical transformations, in so far as nitrogen is in these cases often kept at the same oxidation level as it is in ammonia, and sulfur at the same oxidation level as it is in hydrogen sulfide. This means that no electron balance has to be taken in consideration concerning nitrogen and sulfur. If necessary this point can be corrected afterwards. To introduce this concept, let us first remember that the building of one new C—C bond between any pair of carbon atoms, previously hydrogenated according to:

$$\geqslant C - H + H \leqslant C - \longrightarrow \ \geqslant C - C \leqslant + 2 \ e^- + 2 \ H^+$$

is a removal of an electron pair, *i.e.* a bielectronic oxidation.

In the same way, the introduction of any new covalent bond between one carbon atom and any heteroatom A, *i.e.* O, N or S, initially supplied in its hydrogenated state, H_2O, H_3N or H_2S, is also a bielectronic oxidation, as, for example:

$$\geqslant C - H + H - A \longrightarrow \ \geqslant C - A + 2 \ e^- + 2 \ H^+$$

Therefore, we may define the oxidation degree of any particular carbon atom in a molecule, which must be known here by its developed formula, by the sum of:

—   the number of covalent bonds between a given carbon atom and hetero-atoms O, N or S;
—   half the number of covalent bonds between this carbon atom and other carbon atoms.

In glucose, the oxidation degrees of all carbon atoms are therefore the following:

$$O=CH–CHOH–CHOH–CHOH–CHOH–CH_2OH$$

$$2.5 \quad 2 \qquad 2 \qquad 2 \qquad 2 \qquad 1.5$$

· The lowest oxidation degree of carbon is reached in methane, where it is zero, which represents the minimum of the scale of oxidation degrees of carbon atoms. The highest value is reached in carbon dioxide where the oxidation degree is 4. The total oxidation degree (TOD) of a given molecule, in reference to methane, is the sum of the oxidation degrees $\Sigma_c ODC$ of all carbon atoms that it contains. Any increase of one unit of oxidation degree is a bielectronic oxidation, any decrease of one unit is a bielectronic reduction. To define globally the level of oxidation of carbon in a given molecule, it is also useful to consider the average value of the oxidation degrees of its carbon atoms:

$$\text{Average Oxidation Degree (AODC)} = \frac{\Sigma_c ODC}{\text{number of C atoms}}$$

In this respect:
—   ethanol, fatty alcohols or polymethylenic chains are at the average oxidation degree 1;
—   saturated fatty acids have average oxidation degrees slightly higher than 1, in correspondence to their longer carbonaceous chain;
—   sugars, acetic acid, lactic acid, alanine are at the average oxidation degree 2;
—   a few biochemical molecules have a AODC between 2 and 4, e.g. pyruvic, malic, oxaloacetic, aspartic acids, serine, nucleic bases.

To calculate the charge balance implied by any conversion of carbonaceous biochemical reagents to products involving the same number of carbon atoms, one has to compare the total oxidation degrees of both groups of components and see if they are not equal to reestablish the balance by adding a number of electrons equal to twice the difference on the most oxidized side; e.g. for the global conversion of glucose to acetate and CO_2:

$$CHO–(CHOH)_4–CH_2OH \longrightarrow 2\,CH_3CO_2H + 2\,CO_2 \qquad + 8\,e^-$$

$$TOD = (7 + 5) = 12 \qquad\qquad TOD = 2\,(3 + 1) + 2 \times 4 = 16$$

Afterwards, the oxygen (and possibly nitrogen or sulfur) balance has to be settled by adding water (or ammonia or hydrogen sulfide) molecules on the side where oxygen (or nitrogen or sulfur) is missing, and the hydrogen balance is finally set by adding protons in the same way on the proper side. As a final control, the charge balance must be right at this stage.

In the given example:

$$C_6H_{12}O_6 + 2 H_2O \longrightarrow 2 CH_3-CO_2H + 2 CO_2 + 8 e^- + 8 H^+$$

2.2.2. Oxidation numbers

The concept of oxidation degree can in fact be used, as here introduced, only when reagents and products involved in biochemical transformations are known by their developed formulas. This is not always the case but we will see now that the electron and mass balances of redox transformations of biochemicals can also be simply settled if one only knows their global formula. For this purpose, we must introduce another way of defining the oxido-reduction state of carbonaceous molecules. It simply consists in extending to carbon the concept of oxidation number, classically used in inorganic chemistry. Let us remind here that the oxidation number of an element in a compound is simply the number of electrons which have to be removed from the atom of the element itself in order to introduce it in the compound in question. Therefore, if we assume here that hydrogen, oxygen and nitrogen atoms can only be exchanged from and to biochemicals respectively at their oxidation numbers $+ 1$, $- 2$ and $- 3$, which is generally the case in biochemistry of carbonaceous substrates (and can possibly be corrected), the total oxidation number (TONC) of carbon atoms in a compound of global formula $C_xH_yO_zN_t$ is:

$$TONC = - (y - 2z - 3t)$$

and the average oxidation number (AONC) of each carbon atom:

$$AONC = - \frac{y - 2z - 3t}{x}$$

In methane, this oxidation number reaches its minimum value at -4. In carbon dioxide, it reaches its maximum value at $+ 4$. In fact, the average oxidation degree AODC and number AONC of carbon atoms in any given biochemical are connected by the simple relationship:

$$AONC = 2 AODC + 4$$

From that stage, to calculate the charge balance implied by any transformation of biochemical reagents to products containing respectively the same number of carbon atoms, one has to compare their TONC and, if they are not equal, to reestablish the balance by adding as many electrons as the difference to the most oxidized side, the following being identical as when using oxidation degrees. The complete reduction till CH_4, H_2O and NH_3 of $C_xH_yO_zN_t$ is then simply:

$$C_xH_yO_zN_t + (4x - y + 2z + 3t)(e^- + H^+) \longrightarrow x\,CH_4\,(g) + z\,H_2O + t\,NH_3$$

In the same way, its complete oxidation to CO_2, H_2O and NH_3 is:

$$C_xH_yO_zN_t + (2x - z)\,H_2O \longrightarrow x\,CO_2(g) + t\,NH_3 + (4x + y - 2z - 3t)(e^- + H^+)$$

2.2.3. Global redox metabolism of biochemicals

Three kinds of global redox reactions play a particularly important role, as we will confirm later on, in biochemistry and in carbonaceous chemistry on the primitive earth:

—   The complete reduction of a carbonaceous substrate by hydrogen, or the reverse oxidation:

$$C_xH_yO_zN_t + (4x - y + 2z + 3t)(e^- + H^+) \rightleftarrows x\,CH_4\,(g) + z\,H_2O + t\,NH_3$$

$$H_2\,(g) \rightleftarrows 2\,H^+ + 2\,e^-$$

$$C_xH_yO_zN_t + \frac{4x - y + 2z + 3t}{2}\,H_2\,(g) \rightleftarrows x\,CH_4\,(g) + z\,H_2O + t\,NH_3$$

—   The complete oxidation of substrates by oxygen which is finally completely reduced to water, or the reverse reactions which globally produce biochemicals on the today's earth. In this case, ammonia also tends, as we will see later on, to be oxidized to nitrogen:

$$C_xH_yO_zN_t + (2x - z)\,H_2O \rightleftarrows x\,CO_2(g) + t\,NH_3 + (4x + y - 2z - 3t)(e^- + H^+)$$

$$(t/2)[2\,NH_3 \rightleftarrows N_2\,(g) + 6\,(e^- + H^+)]$$

$$(1/4)(4x + y - 2z - 3t)[O_2\,(g) + 4\,H^+ + 4\,e^- \rightleftarrows 2\,H_2O]$$

$$C_xH_yO_zN_t + \frac{2x + (1/2)y - z}{2}\,O_2\,(g) \rightleftarrows x\,CO_2\,(g) + (1/2)y\,H_2O + (1/2)t\,N_2(g)$$

The complete oxidation of a part of a substrate to CO_2 can supply the electrons required for the complete reduction of its other part to CH_4, which results in the complete disproportionation of this substrate, as:

$$(1/8) (4x - y + 2z + 3t) [(C_xH_yO_zN_t + (2x - z) H_2O \rightleftarrows$$

$$\rightleftarrows x CO_2 (g) + t NH_3 + (4x + y - 2z - 3t) (e^- + H^+)]$$

$$(1/8) (4x + y - 2z - 3t) [C_xH_yO_zN_t + (4x - y + 2z + 3t) (e^- + H^+) \rightleftarrows$$

$$\rightleftarrows x CH_4 (g) + z H_2O + t NH_3]$$

$$C_xH_yO_zN_t + (1/4) (4x - y - 2z + 3t) H_2O \rightleftarrows$$

$$\rightleftarrows (1/8) (4x + y - 2z - 3t) CH_4 (g) + (1/8) (4x - y + 2z + 3t) CO_2 (g) + t NH_3$$

In fact, in any anaerobic transformation of biochemicals, since no external electron exchangers are involved, some carbon atoms must increase their oxidation degree and some others must decrease it. In lactic fermentation from glucose:

$$O=CH-(CHOH)_4-CH_2OH \longrightarrow 2 CO_2H-CHOH-CH_3$$

ODC $2.5 + 2 \times 4 + 1.5 \longrightarrow 2 \times (3.5 + 2 + 0.5)$

the oxidized carbon atoms and the reduced ones remain linked in the same molecules. But, in ethanolic fermentation, the most oxidized part of the molecule comes off because of the instability of C—C bonds between highly oxidized carbon atoms:

$$O=CH - (CHOH)_4 - CH_2OH \longrightarrow 2 CO_2 + 2CH_2OH - CH_3$$

ODC $2.5 + 2 \times 4 + 1.5 \longrightarrow 2 \times 4 + 2(1.5 + 0.5)$

which altogether produces molecules with average oxidation degrees respectively higher and lower than the AODC of the initial reagent.

3.   Energy balances of redox reactions

The energy balance ΔG_r of any redox reaction stoichiometrically defined, including the developed formula of its components, can be settled exactly like any other chemical process involving neutral species, ions and eventually water, the solvent, itself, e.g. for:

$$a Ox_1 + b Red_2 + q H_2O \rightleftarrows c Red_1 + d Ox_2 + p H^+$$

$$\Delta G_r^\circ = c \Delta G_f^\circ(Red_1) + d \Delta G_f^\circ(Ox_2) + p \Delta G_f^\circ(H^+) - a \Delta G_f^\circ(Ox_1) +$$

$$- b \Delta G_f^\circ(Red_2) - q \Delta G_f^\circ(H_2O)$$

where all ΔG_f° refer to standard free enthalpies of formation of corresponding components from the elements they contains taken in their most stable states. In the proton reference system, ΔG_f° (H$^+$) is taken equal to zero. The standard free enthalpy of water formation ΔG_f° (H$_2$O) is constant at 25 °C at -56.69 kcal mole^{-1}. All other free enthalpies of formation are given according to choices made, either in solution at unit concentration, or in the gaseous state at unit pressure or in the solid state. The same relationship holds for enthalpies H. The equilibrium constant is then given by:

$$RT \log K = -\Delta G_r^{\circ}$$

where:

$$K = \frac{[\text{Red}_1]^c \, [\text{Ox}_2]^d \, [\text{H}^+]^p}{[\text{Ox}_1]^a \, [\text{Red}_2]^b}$$

if all components quantities are expressed by their molarities. The solvent H$_2$O must not be mentioned in the fraction. The molarities must be replaced by partial pressures if the reference quantities of components are given in the gaseous state. They must be omitted if the components are insoluble or maintained at saturation.

3.1. Energy balances of global redox metabolism

The energy balances of global redox metabolisms previously defined by their stoichiometry can be calculated from these relationships. A first result of such evaluations is that all complete hydrogenations of any biochemical to methane are exoenergetic. They produce between 8 and 15 kcal mole^{-1} per C, N or O atom contained in the molecule. This means that:

—   on primitive earth all syntheses of biochemicals from methane, ammonia and water could not be fulfilled without some kind of physical energy supply;
—   the biological complete reduction of any biochemical by hydrogen is always energetically possible. In fact it is known experimentally with some kind of thermophilic methanogenic bacteria.

Secondly, all complete oxygenations of biochemicals are also exoenergetic, and they generally produce three to five times more energy than the hydrogenation of the same compound. This means, as it is well known experimentally, that:

—   all complete combustions of biochemicals are energetically feasible.
—   on the today's earth, the production of biochemicals from CO$_2$, N$_2$ and water always require energy supply, which is performed through photosynthesis.

Finally, the standard free enthalpies of all complete disproportionations have also negative values for any organic compound, except the element carbon itself and probably some polybenzenic compounds (Table 1). Such so-called disproportionations are fulfilled until both complete oxidation to CO_2 and complete reduction to CH_4 of carbon atoms in methanogenic digestion. From this point of view, the biological production of methane and carbon dioxide, *i.e.* respectively the lowest and highest oxidation degrees of carbon, from any biochemical corresponding to an average oxidation level of carbon intermediate between these two extremes, is merely a complete redox disproportionation of the carbonaceous part of this biological molecule.

4.   Energetics of electron exchanges with redox couples

4.1.   Redox potential

The chemical potential of exchangeable electrons with redox couples in solutions is currently defined by the oxido-reduction (redox) potential U corresponding to the free enthalpy change occurring when $1/F$ equivalents of electrons are supplied by the solution. The zero level of this redox potential is usually defined for a H^+ solution of molar activity saturated with gaseous H_2 under 1 atm.

The NERNST formula, which is a mere consequence of both principles of energetics for the case of ideally diluted solutions, defines the relationship between the redox potential, U, of a solution and the nature and concentrations of its components for any solution containing a given redox couple. In the most general case of a redox couple obeying the stoichiometry:

$$a \text{ Ox} + n \text{ e}^- + p \text{ H}^+ + r \text{ M} \rightleftharpoons b \text{ Red} + q \text{ H}_2\text{O}$$

this formula can be written:

$$U = U^\circ + \frac{R\,T}{n\,F} \ln \frac{[\text{Ox}]^a\,[\text{H}^+]^p\,[\text{M}]^r}{[\text{Red}]^b}$$

if the references states of the dissolved components and water are respectively molar activities and unit molar fraction, *i.e.* pure water. If, when writing the redox couple stoichiometry, the symbol (g) is placed behind Ox, Red or M, the components concerned are considered to be exchanged in the gaseous state, and their molarities are replaced by their partial pressures in the ln expression as well as in the definition and calculation of U° from standard free enthalpies of formation. If, similarly, the

Table 1. – Energy balances of disproportionation reaction of biochemicals to methane and carbone dioxide

Disproportionation	kcal mole^{-1}			
	$\Delta G°$	$\Delta G°/C$	$\Delta H°$	$\Delta H°/C$
2 Ethanol $\longrightarrow$ 3 CH$_4$ (g) + CO$_2$ (g)	– 47,14	– 11.78	– 15	– 3.75
4(–CH$_2$–) + 2 H$_2$O $\longrightarrow$ 3 CH$_4$ (g) + CO$_2$ (g)	– 9.1	– 2.27	+ 8.64	+ 2.16
Acetic acid $\longrightarrow$ CH$_4$ (g) + CO$_2$ (g)	– 12.6	– 6.30	+ 4.46	+ 2.23
β D-glucose $\longrightarrow$ 3 CH$_4$ (g) + 3 CO$_2$ (g)	– 99.98	– 16.67	– 30.28	– 5.05
4 L-serine + 2 H$_2$O $\longrightarrow$ 5 CH$_4$ (g) + 7 CO$_2$ (g) + 4 NH$_3$	– 146.22	– 12.18	+ 5.96	+ 0.50
4 glycine + 2 H$_2$O $\longrightarrow$ 3 CH$_4$ (g) + 5 CO$_2$ (g) + 4 NH$_3$	– 65.38	– 8.17	+ 40.76	+ 5.10
4 adenine + 30 H$_2$O $\longrightarrow$ 5 CH$_4$ (g) + 15 CO$_2$ (g) + 20 NH$_3$	– 182.2	– 9.11		

Among carbonaceous derivatives only carbon itself and probably some polyaromatic hydrocarbons present a positive standard free enthalpy balance for their disproportionation to methane and carbon dioxide.

symbol (s) appears, the components concerned are considered to be exchanged in their pure solid state and they are omitted in the ln term, but taken into account as such in the definition and calculation of $U°$. At 25 °C,

$$(RT/F) \log(10) = 2.3 \, RT/F = 0.05916 \text{ V}$$

In any pH region where the major protonation state of every component of a redox couple remains unchanged, the NERNST formula is usually written:

$$U = U^{o\prime} + \frac{2.3 \, R \, T}{n \, F} \log \frac{[Ox]^a [M]^r}{[Red]^b}$$

where

$$U^{o\prime} = U^\circ - \frac{2.3 \, R \, T}{F} \frac{p}{n} \, (\text{pH})$$

$U°$ is the standard redox potential of the couple concerned; it corresponds to the redox potential of a solution where $[Ox] = [Red] = [M] = [H^+] = 1 \, M$, $i.e.$ at pH = 0. $U^{o\prime}$ is currently called the apparent standard redox potential at any defined pH. It corresponds to the redox potential of a solution where $[Ox] = [Red] = [M] = 1 \, M$ buffered at the concerned pH. Diagrams for any given redox couple, $U^{o\prime}$ versus pH, which represent the variations of $U^{o\prime}$ against pH, considering all possible changes of the dominant protonation states of its components, are easily constructed (see $e.g.$ Ref. 1 and 2). They offer an easy way of appreciating at a glance the redox and acid-base properties of any redox couple, $e.g.$ concerning the range of predominance of its oxidized and reduced forms in all their protonation states or its reaction capacity with any other given redox couple.

4.2. Standard free enthalpy balance of redox reactions

For any redox reaction, such as:

$$n_2 a_1 \, Ox_1 + n_1 b_2 \, Red_2 + (n_2 p_1 - n_1 p_2) \, H^+ \rightleftharpoons$$

$$\rightleftharpoons n_2 b_1 \, Red_1 + n_1 a_2 \, Ox_2 + (n_2 q_1 - n_1 q_2) \, H_2O; \qquad \Delta G°(1/2)$$

where $\Delta G°(1/2)$ represents the standard free enthalpy balance for the reaction of Ox_1 with Red_2, the equilibrium condition is: $U_1 = U_2$, where U_1 and U_2 are the redox potentials calculated for the two redox couples Ox_1/Red_1 and Ox_2/Red_2 respectively. From this relationship, the classical mass action law can be deduced:

$$\frac{[Red_1]^{n_2 b_1} [Ox_2]^{n_1 a_2}}{[Ox_1]^{n_2 a_1} [Red_2]^{n_1 b_2} [H^+]^{(n_2 p_1 - n_1 p_2)}} = K(1/2)$$

where $K(1/2)$ represents the mass action constant for the oxydation of Red_2 by Ox_1 with $RT \ln K(1/2) = -\Delta G^\circ (1/2)$ and $\Delta G^\circ (1/2) = -n_1 n_2 F(U_1^\circ - U_2^\circ)$.

For an usual type reaction:

$$Ox_1 + Red_2 + \ldots \rightleftarrows Red_1 + Ox_2 + \ldots$$

where n electrons are exchanged from Red_2 to Ox_1, when starting from equal quantities of Ox_1 and Red_2 at zero time in the absence of products, the degree of advancement at equilibrium is given by

— about 10^{-a} if a is defined by $U_1^{\circ\prime} - U_2^{\circ\prime} = -(2 \times 0.06 \frac{a}{n})$ (V)

— 0.5 if $U_1^{\circ\prime} = U_2^{\circ\prime}$

— about $(1 - 10^{-b})$ if b is defined by $U_1^{\circ\prime} - U_2^{\circ\prime} = +(2 \times 0.06 \frac{b}{n})$ (V)

4.3. Associations of redox couples

Several simple associations of redox couples are of particular importance in biochemistry.

4.3.1. Successive redox couples, disproportionation reactions

When the oxidant of a first redox couple,

$$Amph + n\, e^- + \ldots \rightleftarrows Red + \ldots; \qquad U^{\circ\prime} (Amph/Red)$$

can be oxidized with the same electron balance:

$$Ox + n\, e^- + \ldots \rightleftarrows Amph + \ldots; \qquad U^{\circ\prime} (Ox/Amph)$$

it is easy to show [3], by considering the additivity of standard free enthalpy balances of reactions between these couples and any other reference one, than the apparent standard potential of the global couple:

$$Ox + 2n\, e^- \rightleftarrows Red + \ldots; \qquad U^{\circ\prime} (Ox/Red)$$

is:

$$U^{\circ\prime}(Ox/Red) = \frac{U^{\circ\prime}(Ox/Amph) + U^{\circ\prime}(Amph/Red)}{2}$$

If, in addition: $U^{\circ\prime}(Amph/Red) > U^{\circ\prime}(Ox/Amph)$ the redox reaction of the so-called ampholytic compound, Amph, on itself, called disproportionation:

$$2 \text{ Amph} + \ldots \rightleftharpoons Ox + Red + \ldots \Delta G^{\circ\prime}$$

is or should be dominantly shifted at equilibrium towards the decomposition of Amph.

It is well known [4] that this is the case of redox potentials of aldehyde/alcohol and carboxylic acid/aldehyde redox couples, which imply that aldehydes should be disproportionated in aqueous solutions if all redox reactions were kinetically possible. In fact, as shown in Fig. 1, it is also the case of all intermediate oxidized states of carbon atoms and also of other metalloids such as oxygen (Fig. 2) and nitrogen and sulfur [1].

4.3.2. Parallel associations of redox couples

Several redox transitions are currently possible between two given oxidation states of organic compounds. This is the case, *e.g.* when different protonation states of the involved oxidized and reduced states occur, such as in:

$$
\begin{array}{l}
CH_3-COOH \xrightarrow[\;-2\,e^- - 2\,H^+\;]{+2\,e^- + 2\,H^+} \\[2em]
+H^+ \Big\updownarrow -H^+ \qquad CH_3-CHO + H_2O \\[2em]
CH_3-COO^- \xrightarrow[\;-2\,e^- - 3\,H^+\;]{+2\,e^- + 3\,H^+}
\end{array}
$$

But some cases, where more complex changes of bonds are involved, are also important for the understanding of several biochemical steps. For instance in the degradative oxidation of β-hydroxyacylcoA or thioesterifying oxidations of alde-

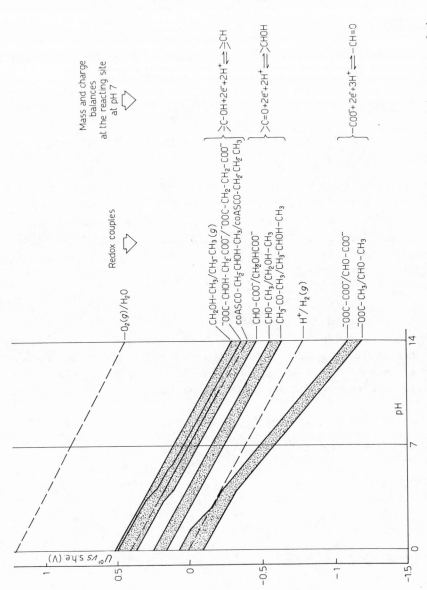

Fig. 1. — $U^{o'}$/pH diagrams for typical bielectronic redox couples involving carbonaceous compounds in aqueous solution.

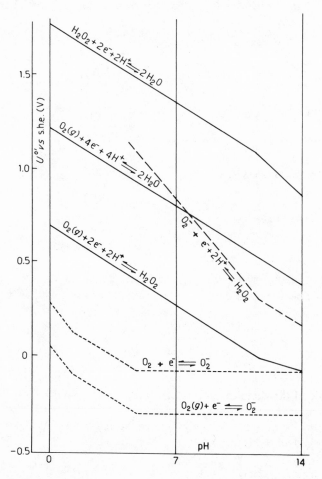

Fig. 2. — $U^{o'}$/pH diagrams for redox couples involving oxygen, hydrogen peroxide, superoxide and water.

hydes, as we will see later on, either the same reductant can produce with the same electron balance two different oxidants, or the same oxidant can be reduced to two different reductants. Even, more generally, sometimes two different reductants at the same total oxidation degree can be oxidized to two different oxidants with the same electron balance. This corresponds e.g. to the case when a free cation such as Fe^{2+} and one of its complexes such as a ferrous cytochrome can be oxidized with the same electron balance, e.g. in this example respectively to Fe^{3+} and ferric cytochrome.

For any such network of process, represented in the most general way by:

$$Ox_1 + n\,e^- \xrightleftharpoons{U^{\circ\prime}_1} Red_1 + \ldots$$

(Other components) $\pm$ $\pm$ (Other components)

$\Delta G^{\circ\prime}_{Ox}$ (downwards) $\Delta G^{\circ\prime}_{Red}$ (downwards)

$$Ox_2 + n\,e^- \xrightleftharpoons{U^{\circ\prime}_2} Red_2 + \ldots$$

involving non-redox processes:

$$Ox_1 + \ldots \rightleftharpoons Ox_2 + \ldots, \qquad \Delta G^{\circ\prime}_{Ox}$$

$$Red_1 + \ldots \rightleftharpoons Red_2 + \ldots, \qquad \Delta G^{\circ\prime}_{Red}$$

it is also easy to show, by considering the additivity of standard free enthalpies of the non-redox reactions and of the redox reactions of both couples with any other reference couple, that

$$n\,F\,(U^{\circ\prime}_1 - U^{\circ\prime}_2) = \Delta G^{\circ\prime}_{Ox} - \Delta G^{\circ\prime}_{Red}$$

5.   Principle of the chemical storage of energy by photosynthesis

Chemically speaking, the energy stored by photosynthesis is oxidoreduction energy. Water is oxidized into oxygen as:

$$(12)[H_2O \rightleftharpoons (1/2)O_2\,(g) + 2\,e^- + 2\,H^+]$$

and carbon dioxide is reduced to glucose, as:

$$6\,CO_2\,(g) + 24\,e^- + 24\,H^+ \rightleftharpoons C_6H_{12}O_6 + 6\,H_2O$$

Both processes correspond to the total oxidoreduction reaction:

$$6\,CO_2\,(g) + 6\,H_2O \rightleftharpoons C_6H_{12}O_6 + 6\,O_2\,(g), \qquad \Delta G^{\circ} = +\,686\ kcal\ mole^{-1}$$

In order to understand how this reaction occurs, we must consider that, in fact, glucose is not the primary reductant formed under the effect of light. Fundamentally, this accumulation of redox energy is based on a very simple principle which has been but oversophisticated by evolution for optimizing the global work of plants, including all their functions. In the leaves and in photosynthetic unicellular organisms, the light energy is first absorbed by pigments and then transferred from pigment (P) to pigment until it finally produces electron excitation of particular photo-receptors:

$$P_f \xrightarrow{\;N_A h\nu\;} P^*, \qquad N_A h\nu_{lum} > \Delta G^\circ_{exc} > N_A h\nu_{abs}$$

where ΔG°_{exc} is the standard free enthalpy of excitation of P_f and ν_{abs} and ν_{lum} represent respectively the minimum absorption wave number and the maximum luminescence wave number of these photoreceptors. The peculiarity of the photo-receptors concerned is to be monoelectronic reductants. If the electron is taken from the fundamental state, this reducing character is described by:

$$P_f \rightleftarrows P^+ + e^-, \qquad U^\circ(P^+/P_f)$$

But this electron elimination can also be performed from the excited state P^*:

$$P^* \rightleftarrows P^+ + e^-, \qquad U^\circ(P^+/P^*)$$

and, when in this case, the standard potential of the redox couple P^+/P^* is very much lower, as it results from the relationship given at in 4.3.2., than the corresponding one for P^+/P_f couple:

$$U^\circ(P^+/P^*) = U^\circ(P^+/P^*) - \frac{\Delta G^\circ_{exc}}{F}$$

This decrease of the standard potential reaches currently more than 1 V for electronic excitations in the visible range of wavelengths. Therefore, P^* can reduce a relatively weak oxidant Ox_1, which could not be reduced by P_f

$$P^* + Ox_1 \longrightarrow P^+ + Red_1$$

This process produces P^+ and the reducing power of P^* is transferred in Red_1 which has to be eliminated by inhibiting the reverse reaction. But P^+ is now a relatively strong oxidant when it is reduced to the fundamental state P_f. In this way, it can oxidize a relatively weak reducer Red_2:

$$P^+ + Red_2 \longrightarrow P_f + Ox_2$$

This produces the corresponding relatively strong oxidant Ox_2 which shall also be eliminated along another path from Red_1. Now P_f can be recycled. The net

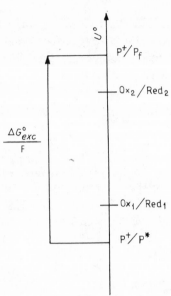

Fig. 3. — Relative values of redox potentials involved in the primary acts of photo-synthesis.

result of these processes:

$$P_f \xrightarrow{\overset{N_A h\nu}{\downarrow} \overset{heat}{\uparrow}} P^*$$

$$P^* + Ox_1 \longrightarrow P^+ + Red_1$$

$$P^+ + Red_2 \longrightarrow P_f + Ox_2$$

$$\overline{Ox_1 + Red_2 \longrightarrow Red_1 + Ox_2 , \qquad \Delta G_r^\circ = -F\,(U_1^\circ - U_2^\circ)}$$

is an accumulation of redox energy ΔG_r° if (Fig. 3):

$$U^\circ\,(P^+/P_f) > U^\circ\,(Ox_1/Red_1) > U^\circ(Ox_2/Red_2) > U^\circ(P^+/P^*)$$

The evolution has simply complicated this very simple principle to the utmost for the best benefit of photoautotrophs, taking into account all chemicals which can be obtained in the presence of water. But without looking more extensively at all these details, notwithstanding their biological importance, we can now go further from the fact that redox energy is obtained from light energy in living systems and look at some conditions which determine the relaxation of this redox energy.

6.   Kinetic criteria for the fulfilment of redox reactions

From any organic compound, many electron-exchanging stoichiometric equations can be written, involving either different levels of oxidation of C-atoms to the total oxidation to CO_2 or different reductions to the total one to CH_4, or even simply different isomers. Considering the apparent standard potentials of all the oxidation processes, it appears, as previously seen, very frequently that they decrease when the involved balance of electrons increases, *i.e.*, that the most complete oxidation processes are energetically the easiest ones. Symmetrically, the most complete reductions are in such cases the energetically easiest ones, *i.e.* occur at the highest potentials. Consequently, all intermediate oxidation levels of carbon should undergo disproportionation if complete equilibrium was reached, and the reduction of the most oxidized state should proceed directly to the most reduced one and reciprocally at potentials equal to the mean value of the potentials of all the intermediate steps. However, it is well known that only redox processes involving a small number of electrons can effectively be observed. In this respect, the experimental facts, as well as logical evidence at the molecular scale lead to the consideration that two general kinetic interdiction rules limit the feasibility of redox processes [3, 5].

Rule 1

No redox process, which should involve at least one couple exchanging more than 2 electrons, is directly feasible at or near its equilibrium state, *i.e.* near the minimal energetic requirements which could make it globally possible. The logical evidence of this rule originates in the fact that relative positioning of the molecules (allowing simultaneous transfer of several electrons doublets by any proper multiple overlapping of orbitals or tunnel effect when the reagents of both involved redox couples are in contact) cannot be conceived according to all the available concepts relating to accessible configurations and electronic structures of molecules.

But experimental evidence shows that a second kinetic rule, which forbids many mono- or bi-electronic redox couples to be implicated in intermolecular electron transfer reactions, must also be considered.

Rule 2

There exists neither experimental evidence nor logical reasons for considering that redox reactions are feasible by *intermolecular electron transfer* at or near their equilibrium conditions when at least one of the two implicated redox couples involves any modification of the covalent topology (or attributive vicinity) of its relatively heavy atoms. In fact, this rule can be considered as a transposition to redox chemistry of the well known FRANCK–CONDON rule currently admitted in spectroscopy. Here it appears that the inertia of electrons is so small in comparison with the relatively large ones of C, O, N or S atoms, that the order of magnitude of the time intervals which are necessary for building or breaking a simple covalent linkage between two such heavy atoms, *i.e.* to change their relative positioning by more than approximately 1 Å, is very much larger than the time intervals involved in the transfer of electrons between two different molecules when they are in the best position for such a process to occur.

Consequently, redox reactions which should imply kinetically forbidden redox processes regarding one of these rules, can proceed only through pathways involving sequences of kinetically permitted steps, each involved step implying its own energetic conditions for taking place. The redox couples involved in redox reactions of these sequences are necessarily restricted to either couples exchanging simply one or two electrons or to couples exchanging one or two electrons together with protons. In the latter case, additional restrictions must be considered in connection with the energetic and kinetic mobility of electrons and protons from or to the reacting oxidant and reductant.

6.1. *Intermolecular electron transfers followed by modifications of the covalent topology of heavy atoms*

The simplest biochemical example of oxidation globally modifying the covalent topology of heavy atoms in a substrate is found in pathways implicating in a sequence a permitted intermolecular electron transfer and a non redox modification of covalent topology of heavy atoms. One very simple example is the oxidation of acyl coA or of succinate. In this latter case, the redox couple which corresponds to the energetically easiest bielectronic oxidation of succinate should introduce an alcoholic group according to:

$$^-OOC-CH_2-CH_2-COO^- + H_2O \rightleftarrows {}^-OOC-CHOH-CH_2-COO^- + 2\,H^+ + 2\,e^-$$

where $U^{o\prime} = +\,0.012$ V(s.h.e.) at pH 7 [4].

But it must be considered as forbidden because of the change of covalent topology it involves between the C(2) of succinate and the O atom from water. In fact, it is biochemically replaced by the sequence of:
—   an allowed intermolecular electron transfer occurring near equilibrium with flavoproteins, which involves the $\alpha\beta$-dehydrogenation:

$$^-OOC-CH_2-CH_2-COO^- \rightleftarrows {}^-OOC-CH = CH-COO^- + 2\,e^- + 2\,H^+$$

where $U^{o\prime} = +\,0.031$ V(s.h.e.) at pH 7 [4]

and needs a higher standard potential than the globally obtained oxidation,
— and, an exoenergetic hydration, but here weakly exoenergetic, of the ethylene
  bond setting the change of covalent topology at a larger time scale:

$$^-OOC-CH=CH-COO^- + H_2O \rightleftharpoons {}^-OOC-CHOH-CH_2-COO^-$$

with $\Delta G^{\circ\prime} = - 0.87$ kcal mole^{-1} [4].

In this sequence, the additional redox standard free enthalpy $\Delta G^{\circ\prime}_{add}$, which is
necessarily involved in the oxidation step because of the kinetic inhibition, in
excess above the standard free enthalpy of the global oxidation:

$$\Delta G^{\circ\prime}_{add} = 2 \times 23.05\ (0.031 - 0.012),\ (kcal\ mole^{-1}) = + 0.87\ kcal\ mole^{-1}$$

is later simply recovered during the hydration process.

But a most important consequence of such a combination of energetic criteria
and kinetic inhibition rules is the opportunity they offer for converting redox
energy into other forms of chemical energy, *e.g.* by chemically coupling the production of a high-energy condensed derivative with the relaxation of redox energy (6).
This is the case, *e.g.* in the oxidation of β-hydroxyacyl coA which occurs in the
catabolism of fatty acid chains, according to:

```
COSco                        COScoA                  COScoA
 |                             |                       |
CH₂    ◄──────┬──────┬──────►  CH₂      + HScoA       CH₃
 |            │      │         |      ──────────►      |
CHOH         ▼      ▼          CO       - HScoA       COScoA
 |        NAD⁺  NADH           |                       |
R                             R                       R
```

In this case, the energetically easiest bielectronic oxidation of the 3-hydroxyl
group should involve (Fig. 4) the degradative hydrolysis of the C_2-C_3 bond, at
potentials lower than those of H^+/H_2 (*g*) couple, according to:

$$R-CHOH-CH_2-COScoA + H_2O \underset{+\ 2\ e^-,\ +\ 3H^+}{\overset{-2\ e^-,\ -3\ H^+}{\rightleftharpoons}} R-CO_2^- + CH_3COScoA;$$

$$U^{\circ\prime}_d$$

But since this process implies changes of the covalent topology of heavy
atoms, related to the incorporation of one oxygen atom and to the degradation

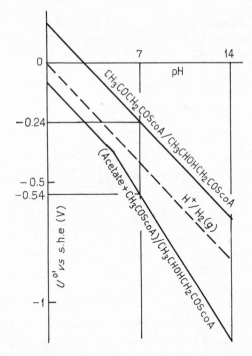

Fig. 4. — Energetics of the overall oxidation of β-hydroxybutyrylcoA into acetate and acetylcoA, as:

$$CH_3CHOHCH_2COScoA + H_2O \rightleftarrows CH_3CO_2H + CH_3COScoA + 2\ e^- + 2\ H^+$$

and of its oxidation into acetoacetylcoA, as:

$$CH_3CHOH\ CH_2COScoA \rightleftarrows CH_3COCH_2COScoA + 2\ e^- + 2\ H^+$$

between C_2 and C_3, it is kinetically blocked. On the other hand, the simpler bielectronic oxidation to the 3-keto derivative:

$$R-CHOH-CH_2-COScoA \underset{+\ 2\ e^-,\ +\ 2\ H^+}{\overset{-\ 2\ e^-,\ -\ 2\ H^+}{\rightleftarrows}} R-CO-CH_2-COScoA$$
$$U^{o\prime}_k \ .$$

is not kinetically forbidden in theory and can be performed by enzyme catalysis with NAD^+. But it occurs at higher potentials and invests an excess of redox standard free enthalpy $\Delta G^{o\prime}_{add}$ such as:

$$\Delta G^{o\prime}_{add} = 2\ F\ (U^{o\prime}_k - U^{o\prime}_d)$$

which is simply the opposite of the apparent standard free enthalpy of the degradative hydrolysis of the C_2–C_3 bond of the β-keto derivative (Fig. 5). This process could further ensure the relaxation of $\Delta G^{o\prime}_{add}$. However, this degradative hydrolysis of the C_2–C_3 bond by H_2O molecules is also kinetically very slow, as are hydrolysis of esters or other acylated condensates. On the contrary, the nucleophiiic attack of $C = O$ by RS^- can occur very much more rapidly as in:

$$
\begin{array}{c}
\qquad\quad H \\
\qquad\quad | \\
O\quad S\text{--}coA \qquad\quad HO\quad ScoA \\
\parallel \qquad\qquad\qquad \backslash\,/ \\
R\text{--}C\text{--}CH_2COScoA \rightleftarrows R\text{--}C\text{--}CH_2\text{--}COScoA \rightleftarrows R\text{--}COScoA + CH_3COScoA
\end{array}
$$

and leads to the recovery of a part of $\Delta G^{o\prime}_{add}$ into a newly formed acyl coA (Fig. 5).

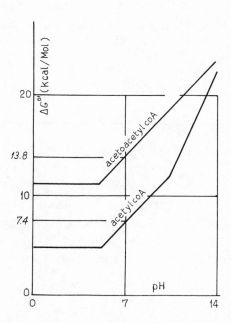

Fig. 5. – Comparison, taking into account pH, of the energies liberated in the hydrolytic degradation of acetoacetylCoA, as:

acetoacetylCoA + H_2O $\rightleftarrows$ acetate + acetylCoA

and used in the condensation leading to acetylcoA from acetate, through:

acetate + HScoA $\rightleftarrows$ acetylcoA + H_2O

6.2. Intermolecular electron transfers anticipated by modifications of the covalent topology of heavy atoms

The modification of the covalent topology of heavy atoms occurring after intermolecular electron transfer in the same pathways, as illustrated before, can alternatively be involved before the electron transfer [6], as much as some energetically and kinetically feasible non redox process allows it. This is the case e.g. with the biochemical oxidation of aldehydes. If considering only bielectronic oxidation, the energetically easiest oxidation of aldehydes should produce acids, according to the equation:

$$\text{R--CHO} + \text{H}_2\text{O} \xrightleftharpoons[+\,2\,e^-,\,+\,2\,H^+]{-\,2\,e^-,\,-\,2\,H^+} \text{R--COOH}$$

at standard apparent potentials $U^{o\prime}_{(ac/al)}$ far much lower than those of H^+/H_2 (g) and nicotinamidic couples (Fig. 6). Nevertheless, this oxidation cannot be achieved near $U^{o\prime}_{(ac/al)}$ or even by H^+ or NAD^+ because it implies the formation of a new C--O bond. However, on the other hand, hemiacetalic forms of oses or hemithioacetals of simple aldehydes are formed at the equilibrium of:

$$\text{R--CH} = \text{O} + \text{HX} \rightleftharpoons \text{R--CH} \overset{\textstyle OH}{\underset{\textstyle X}{\big\langle}} \quad ; \qquad \Delta G^{o}_{(hemi)}$$

where HX is either an alcohol group (conveniently located in the same molecule as the aldehyde one, as in sugar) or a thiol RSH, with standard free enthalpies, $\Delta G^{o}_{(hemi)}$, which are near zero or even slightly negative ($\Delta G^{o}_{(hemi)} = -1.8$ kcal mole^{-1} for the addition of C_2H_5SH on CH_3CHO) and not dependent on pH below the pK_a of XH. Consequently, the oxidation by intermolecular electron transfers of these hemiacetalic forms becomes kinetically possible, since it does not involve any more modifications of the covalent topology of heavy atoms, according to the stoichiometrically simpler oxidation step:

$$\text{R--CH} \overset{\textstyle OH}{\underset{\textstyle X}{\big\langle}} \xrightleftharpoons[+\,2\,e^-,\,+\,2\,H^+]{-\,2\,e^-,\,-\,2\,H^+} \text{R--C} \overset{\textstyle O}{\underset{\textstyle X}{\big\langle}}$$

This now produces a condensed derivative of the acid otherwise obtained, and consequently occurs at apparent standard potential $U^{o\prime}_{(cond/hemi)}$ equal to:

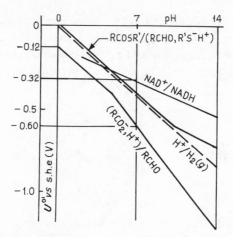

Fig. 6. — Energetics of the redox processes:

$$R\ COO^- + 2\ e^- + 3\ H^+ \rightleftarrows R\ CHO + H_2O$$

and $\quad R\ COSR' + 2\ e^- + 2\ H^+ \rightleftarrows R\ CH(OH)\ SR' \rightleftarrows R\ CHO + HSR'$

$$U^{o\prime}_{(cond/hemi)} = U^{o\prime}_{(ac/al)} + \frac{\Delta G^{o\prime}_{cond} - \Delta G^{o\prime}_{hemi}}{2\ F}$$

where $\Delta G^{o\prime}_{cond}$ is the positive apparent free enthalpy of condensation of the condensation of the condensate R—CO—X. In fact, $\Delta G^{o\prime}_{(cond)}$ being more positive than $\Delta G^{o\prime}_{(hemi)}$, because the R—CO residue is more oxidized than the R—CHOH one, the redox step involved in this pathway occurs at higher potentials than the standard potential of the $H^+/H_2\,(g)$ couple and stores the condensation energy of R—CO—X (Fig. 6).

6.3. Redox reactions occurring through molecular association followed by intramolecular electron shift

The last kind of pathway for performing a redox reaction, which globally modifies the covalent topology of heavy atoms between involved reagents and products without encountering one of the forbidding rules involves successively:

— the non redox formation of one single molecule from the reacting oxidant and reductant;

— an intramolecular shift of electrons in the formed molecule, which evidently occurs without any change of covalent topology of heavy atoms in the single molecule obtained;

— which is, or is not, followed by the breakdown of this transformed intermediate into products respectively reduced and oxidized.

A most classical example of such a sequence occurring enzymatically at its global equilibrium and concluded by a dissociation is the aminating reduction of α-ketoacids by pyridoxamine involved in transamination. Here, a weakly exergonic condensation first occurs from the oxidant α-ketoacid and the reductant pyridoxamine, as:

$$
\begin{array}{c}
R \\
| \\
C=O + H_2N-CH_2-Ar \\
| \\
CO_2^-
\end{array}
\quad
\underset{- H_2O}{\overset{+ H_2O}{\rightleftharpoons}}
\quad
\begin{array}{c}
R \\
| \\
C=N-CH_2-Ar \\
| \\
CO_2^-
\end{array}
$$

forming a single SCHIFF base molecule, wherein an electron shift arises without any change in the covalent topology of heavy atoms, as:

$$
\begin{array}{c}
R \\
| \\
C=N-CH_2-Ar \\
| \\
CO_2^-
\end{array}
\quad \rightleftharpoons \quad
\begin{array}{c}
R \\
| \\
HC-N=CH-Ar \\
| \\
CO_2^-
\end{array}
$$

This tautomery of the SCHIFF base is finally completed here by hydrolysis of the new imino group, as:

$$
\begin{array}{c}
R \\
| \\
HC-N=CH-Ar \\
| \\
CO_2^-
\end{array}
\quad
\underset{- H_2O}{\overset{+ H_2O}{\rightleftharpoons}}
\quad
\begin{array}{c}
R \\
| \\
HC-NH_2 + O=CH-Ar \\
| \\
CO_2^-
\end{array}
$$

In some cases, the same principle is applied from simpler reagents without any final separation into two products. Particularly typical in this respect is the case of far-from-equilibrium hydroxylations of initially non oxidized C atoms, which can even occur non enzymatically in sensible positions of aromatic structures [7, 8]. Such reactions appear to be driven by any process having produced $HO^{\bullet}$ radicals in a previous step, e.g. from reduction of dissolved O_2 to H_2O_2 followed by any catalytic dissociation of the peroxide bond, such as we will see later in Section 7. Such radicals where oxygen is at the oxidation number -1 can attack the RH bond, as:

$$RH + {}^{\bullet}OH \longrightarrow R^{\bullet} + H_2O$$

followed by:

$$R^{\bullet} + {}^{\bullet}OH \longrightarrow ROH$$

or:

$$R^{\bullet} + HOOH \longrightarrow ROH + {}^{\bullet}OH$$

finally resulting in the global oxidation:

$$RH + H_2O \longrightarrow ROH + 2\,e^- + 2\,H^+$$

$$H_2O_2 + 2\,e^- + 2\,H^+ \longrightarrow 2\,H_2O$$

i.e.

$$RH + H_2O_2 \longrightarrow ROH + H_2O$$

which globally performs a redox reaction implying modifications of covalent topologies of both reagents, without involving at any time any intermolecular electron transfer, forbidden because it proceeds too fast to be simultaneous with a change in the covalent vicinity between relatively heavy and mechanically inert atoms.

7.   The oxygen reduction in aqueous media

Now I would like to develop a similar analysis for the case of biochemical oxygen reduction. In fact, it will lead us to very exciting new insights on the mechanism of at least a part of mitochondrial oxidative phosphorylations [9]. First of all, we have to realize that it is all nonsense to blindly admit, as is done in nearly all analyses of biological facts including many biochemistry courses, that oxygen is reducible by the electron transfer chain directly to water, as:

$$O_2\,(g) + 4\,e^- + 4\,H^+ \rightleftarrows 2\,H_2O$$

at or near the equilibrium potential of this redox couple, at 25 °C

$$U(V/s.h.e.) = 1.228 - 0.0591\ pH + (0.0591/4)\log p_{O_2}$$

since it involves 4 electrons and modifies the covalent topology of oxygen atoms. Under the best conditions, kinetically and energetically speaking, this reduction can involve two electrons without modifying the covalent topology of oxygen atoms, *i.e.* produce hydrogen peroxide:

$$O_2\,(g) + 2\,e^- + 2\,H^+ \rightleftharpoons H_2O_2$$

at or near:

$$U(\text{V/s.h.e.}) = 0.682 - 0.0591\ \text{pH} + \frac{0.0591}{2}\ \log \frac{p_{O_2}}{[H_2O_2]}$$

or taking into account the oxygen solubility in water:

$$U = 0.797 - 0.0591\ \text{pH} + \frac{0.0591}{2}\ \log \frac{[O_2]}{[H_2O_2]}$$

In fact, simpler stoichiometries could also be kinetically considered [8, 10] such as the production of superoxide:

$$O_2 + e^- \rightleftharpoons O_2^-$$

but they are only feasible (Fig. 2) at lower potentials and can be biochemically short-circuited thanks to some catalytical effects on the disproportionation of O_2^-. This potential reasonably corresponds to the reduction by cytochrome oxidase which reversibly transfers its electrons at pH 7 at standard potential around 0.27 V/(s.h.e.), since at 90 % reduction of oxygen to H_2O_2 :

$$U = 0.797 - (0.0591 \times 7) - (0.0591/2) = 0.35\ \text{V/(s.h.e.)}$$

In this case, the disproportionation of superoxide can be considered sufficiently fast to allow the reduction to proceed until H_2O_2 at its equilibrium potential. On the contrary, the rate of disproportionation of H_2O_2 is considered here, according to our rules, as being relatively so small that the potential of reduction of O_2 into H_2O_2 is not noticeably increased by any decrease in the concentration of hydrogen peroxide which could be due to this disproportionation. Consequently, if mitochondria are supplied with electrons from an equimolar mixture of *e.g.* malate and oxaloacetate [$U^{o\prime} = -0.166$ V(s.h.e.) at pH 7], the total free enthalpy which is evolved in their electron transfer chain by the transfer or two electrons to oxygen at its half reduction is only:

$$\Delta G^{o\prime} = 2\ F\ (0.383 + 0.166) = -25.3\ \text{kcal}$$

On the other hand, it is known that between 12 and 16 kcal are necessary for producing one ATP molecule from ADP and orthophosphate in the range of cellular concentrations of these substrates [11], because of the concentration effect due to the reduction in the number of dissolved substrates that this reduction implies.

This means that the amount of energy evolved in the electron transfer chain is at the best of its use hardly enough for producing 2 ATP from the oxidation of malate by dissolved oxygen. Since it is experimentally well known that between 2.5 and 3 ATP are produced under these conditions when one mole of lactate is oxidized, the question is now: wherefrom does the complementary amount of energy come out?

Here, we must obviously look for the place where the free enthalpy corresponding to the difference between the standard potentials of oxygen/water and oxygen/hydrogen peroxide redox couples is located. The answer is plain: it is accumulated in the obtained hydrogen peroxide molecule, since its disproportionation, as:

$$H_2O_2 \longrightarrow (1/2)O_2\,(g) + H_2O$$

which results from the redox reaction between the redox couples:

$$\frac{1}{2}\,[H_2O_2 + 2\,e^- + 2\,H^+ \longrightarrow 2\,H_2O] \qquad\qquad U^{\circ\prime} = 1.774 \text{ V(s.h.e.)}$$

and

$$\frac{1}{2}\,[H_2O_2 \longrightarrow \frac{1}{2}O_2\,(g) + 2\,e^- + 2\,H^+]\; U^{\circ\prime} = 0.682 \text{ V(s.h.e.)}$$

$$\overline{\qquad\qquad H_2O_2 \longrightarrow \frac{1}{2}O_2\,(g) + H_2O \qquad\qquad}$$

can now evolve at any pH in the biological range with a $\Delta G_d^{\circ\prime}$ equal to:

$$\Delta G_d^{\circ\prime} = -\,F\,(1.774 - 0.682) = -\,25.2 \text{ kcal}$$

each time two electrons are transferred from malate to oxygen, which precisely corresponds to the standard free enthalpy lost because oxygen is only reducible to hydrogen peroxide instead of water.

Fortunately, in some respects, the redox couple $H_2O_2/2H_2O$ is also kinetically forbidden because of the second rule. Firstly in such a way as hydrogen peroxide cannot simply oxidize all cell components, just as its very high standard potential as an oxidant could permit it to do so if its reduction was not kinetically forbidden. But also, in such a way we must now look more closely at the mechanism of this disproportionation, to understand how it can globally proceed in spite of the second forbidding rule and also in order to examine if a part of the energy evolved could not be easily transferred at this occasion to orthophosphate to allow it to condense with ADP.

Since the difficult part of this disproportionation lies in the reduction of hydrogen peroxide to water because the change of covalent vicinity between both oxygen atoms that it implies cannot occur in the very short time necessary for an electron to be taken out from another molecule, let us look, as we did it previously with regard to the biochemistry of carbonaceous compounds, at how electron transfers and changes of covalent topology could be disconnected here. The simple dissociation of H_2O_2 in two hydroxyl radicals is here impossible because of its too high positive free enthalpy balance:

$$H_2O_2 \longrightarrow 2\ {}^\bullet OH\ ; \qquad \Delta G^\circ = +\ 48.5\ \text{kcal}$$

But a coupling of this process with the oxidation of a monoelectronic reducer, such as a ferrous complex FeX^{2+} by one of the produced radicals is possible through the mechanism:

$$HO-OH + X\,Fe^{2+} \longrightarrow HO-O\,FeX^+ + H^+$$

$$HO-O\,FeX^+ \longrightarrow HO^\bullet + {}^\bullet O\,FeX^+$$

$$-2\,H^+ \updownarrow + 2\,H^+$$

$$X\,Fe^{3+} + H_2O$$

$$\overline{H_2O_2 + X\,Fe^{2+} + H^+ \longrightarrow HO^\bullet + H_2O + X\,Fe^{3+}\ ; \qquad \Delta G^{\circ\prime}_c}$$

which is energetically equivalent to the sum of:

$$H_2O_2 \longrightarrow 2\,HO^\bullet\ ; \qquad +\ 48.5\ \text{kcal}$$

and

$$OH^\bullet + X\,Fe^{2+} + H^+ \longrightarrow H_2O + X\,Fe^{3+}\ ; \qquad \Delta G^{\circ\prime}_r$$

Consequently, the standard free enthalpy balance $\Delta G^{\circ\prime}_c$ of this global process will be near zero, which allows this reaction to occur easily, if:

$$48.5\ \text{kcal} = -\ \Delta G^{\circ\prime}_r = F\,[U^{\circ\prime}\,(OH^\bullet/H_2O) - U^\circ\,(X\,Fe^{3+}\,/\,X\,Fe^{2+})]$$

If we adopt here the value of $+\ 2.4$ V(s.h.e.) for $U^{\circ\prime}({}^\bullet OH/H_2O)$ at pH 7 [1], it means that the apparent potential of the ferric/ferrous complex able to fulfill the catalasic role should stay at most around $+\ 0.3$ V(s.h.e.) which enables a lot of hemic electron exchangers to be good candidates for such a function.

In order to continue with this catalytic disproportionation, we must now look at the reduction to water of the other hydroxyl radicals produced. Here no kinetic limitations at all could forbid the monoelectronic transfer:

$$^{\bullet}OH + X'Fe^{2+} + H^+ \longrightarrow H_2O + X'Fe^{3+}$$

to happen with a largely negative free enthalpy balance, *i.e.* at pH 7:

$$\Delta G^{\circ \prime} = - F\ [2.4 - U^{\circ \prime}\ (X'\ Fe^{3+}/XFe^{2+})]$$

Finally, it remains to regenerate both produced ferric complexes through:

$$H_2O_2 + 2\ (X\ or\ X')\ Fe^{3+} \longrightarrow O_2 + 2\ H^+ + 2\ (X\ or\ X')\ Fe^{2+}$$

for globally obtaining the disproportionation of two molecules of hydrogen peroxide without having at any place transgressed the rule which forbids any change of covalent topology between heavy atoms during the time electrons are being transferred from one molecule to another one. This implies a lower limit for the standard potential of ferric/ferrous redox couples:

$$U^{\circ \prime}\ [(X\ or\ X')Fe^{3+}\ /\ (X\ or\ X')Fe^{2+})] > 0.38\ V(s.h.e.)$$

Taking into account the upper limit of the same potential previously given, we obtain:

$$U^{\circ \prime}[(X\ or\ X')\ Fe^{3+}\ /\ (X\ or\ X')Fe^{2+}\ ] \sim 0.3\ to\ 0.38\ V(s.h.e.)$$

If we now come back to the problem of ATP production, it obviously appears that the best candidate as a coupling process involved in this succession of steps is the reduction of the produced hydroxyl radical by $X'Fe^{2+}$ because it is the only largely exoenergetic step. On this respect, the activation of an orthophosphate can be easily achieved if a part of the complexation of the involved ferrous ion is insured, at a slightly exoenergetic equilibrium, by orthophosphate. Then, the oxidation should result in the production of the same complex of the ferric state, again because of the kinetic rule which here forbids the coordinance bond between orthophosphate and iron to be broken during the time necessary for the electron to be transferred, and it can energetically do so since the standard potential of OH/H_2O redox couple is fairly high. Finally, if the free enthalpy of dissociation of this ferric complex is now sufficiently negative, it could transfer its phosphate group to ADP under an activated form, *e.g.* as a metaphosphate ion (Fig. 7).

8.   Conclusion

As a matter of fact, the main reason why the chemiosmotic theory completely overwhelmed [12] chemical theories of oxidative phosphorylation until now

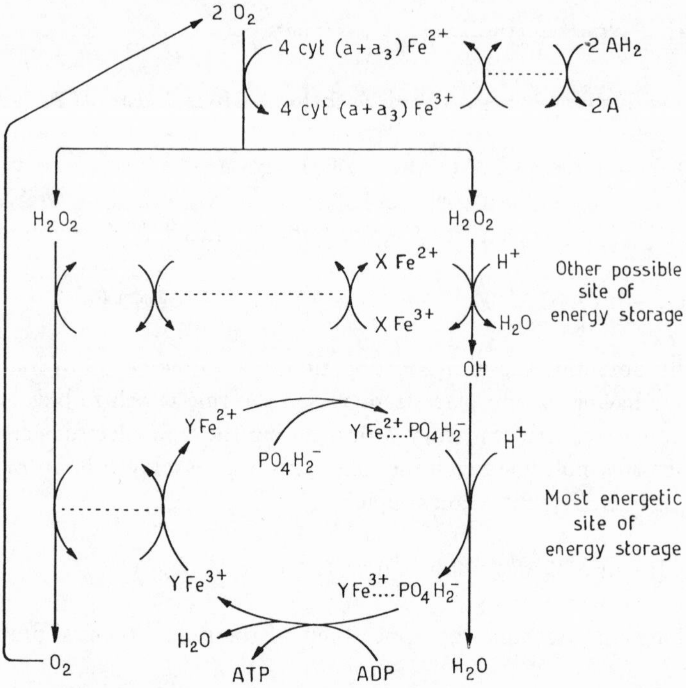

Fig. 7. — Proposed pathway for the chemical conversion of disproportionation energy of H_2O_2 to ATP condensation energy.

lies in the absence of argumented proposals to define the energy rich chemical intermediates implied by these theories. On the contrary, the pH gradient proposed as an intermediate energy storage step by the chemiosmotic theory [13] is a clear electrochemical evidence.

Now, the situation is partially reversed. It plainly appears that hydrogen peroxide is an excellent candidate as an energy-rich intermediate of a phoenix-like chemical theory, and that a lot of hemic electron exchangers, among which catalase and several cytochromes are good candidates for playing the role of catalasic and energy-coupling tools. Moreover the complete mechanism proposed here constitutes a very simple application of general kinetic rules in redox chemistry, which also underlies, as shown previously, other conversions of redox energy to condensation energy by substrate oxidations. Faced with that, the mechanisms of coupling the chemiosmotic pH gradient to ATP production remains at present undefined on energetic grounds.

In conclusion, it appears appropriate to spend as much time and effort as necessary in experimentally testing all possible applications of this chemical coupling mechanism with biological systems and models [14] as has been done for chemiosmotic theory. And to wait for the results to allow supporters of both theories to fight again.

References

[1]   M. POURBAIX, *Atlas d'équilibres électrochimiques à 25 °C,* Gauthiers Villard, Paris (1963).

[2]   R. WURMSER, in *Traité de biochimie générale,* Paris (1967) Tome II, pp. 36-46.

[3]   R. BUVET, *Experientia* **36,** 1254 (1980).

[4]   All numerical data cited are taken from: * *Handbook of Chemistry and Physics* The Chemical Rubber Co, Cleveland, Ohio (1968)
      * *Handbook of Biochemistry,* 2nd ed., The Chemical Rubber Co, Cleveland, Ohio (1970)
      * T.E. BARMAN *Enzyme Handbook,* Springer-Verlag, Berlin (1969).

[5]   R. BUVET, *Critères énergétiques de déroulement des processus de transfers d'électrons,* in *La bioconversion de l'énergie solaire,* Masson, Paris (1981), pp. 125-153.

[6]   R. BUVET, *Energetics of coupled biochemical processes and of their chemical models,* in *Living Systems as Energy Converters,* R. BUVET et. al. (Editors), North Holland, Amsterdam (1977), pp. 21-39.

[7]   S. ISHIMITSU, S. FUJIMOTO and A. OHARA, *Chem. Pharm. Bull. Jpn.,* **25**(3), 471 (1977).

[8]   R. BUVET and L. LE PORT, *Electroch. Acta* **25,** 97 (1979).

[9]   R. BUVET, *Energetics of the chemical coupling between ATP production and oxygen reduction, comparison with electrochemical data,* to be published in Bioelectrochemistry and Bioenergetics.

[10] B.H.J. BIELSKY, *Photochem. Photobiol.* **28,** 645 (1978).

[11] A.L. LEHNINGER, *Biochemistry,* Second Edition, (1975), p. 391.

[12] E.C. SLATER, *Biological membranes as energy transducers,* in *Living Systems as Energy Converters,* R. BUVET et al. (Editors), North Holland, Amsterdam (1977), pp. 221-227.

[13] P. MITCHELL, *Science* **206,** 1148-1159.

[14] M. VUILLAUME, R. CALVAYRAC and M. BEST-BELPOMME, *Biol. Cellul.,* **35,** 71 (1979).

Supplementary useful readings

G. MILAZZO, *Electrochemistry*, Elsevier Publishing Company, Amsterdam, London, New York (1963).

A.L. LEHNINGER, *Biochemistry*, North Holland, Amsterdam, Second Edition (1975).

R. BUVET, M. ALLEN and J.P. MASSUE (Editors), *Living Systems as Energy Converters*, North Holland Publ., Amsterdam (1977).

R. BUVET, *Energetic Structure of Metabolism*, in G. Milazzo (Editor) *Topics in Bioelectrochemistry and Bioenergetics*, J. Wiley, London, New York, (1976) Vol. 1, pp. 105-177.

G. GAVACH (Editor), *La Bioconversion de l'Energie Solaire*, Masson, Paris (1981).

PHOTOSYNTHESIS – SELECTED TOPICS

HELMUT METZNER

Institute of Chemical Plant Physiology
University of Tübingen – 7400 Tübingen 1, G.F.R.

Contents

This short introduction into photosynthesis will not comprise the whole field, but should be taken as a (subjective) selection of chapters which might be the most interesting with respect to bioelectrochemistry.

1.   Historical introduction

The very first test which may be regarded as a photosynthesis experiment dates back to the beginning of the 17th century. It was the Flemish physician JAN VAN HELMONT in Brussels who performed the first quantitative measurement of plant growth. Since the time of ARISTOTLE everybody was convinced that plants take all their food from the soil. To check this assumption VAN HELMONT planted a willow of four pounds weight and irrigated it for five years. Within this time he obtained a small tree which weighted nearly 150 pounds. At the same time the soil had lost just a few grams. So VAN HELMONT concluded that plants take up part of their matter from water.

Nearly two centuries had passed, when the English clergyman JOSEPH PRIESTLEY experimented with different gases. This was still before the decisive studies of LAVOISIER, and so PRIESTLEY's letters contain a very curious terminology. He first realized that mice which were included into a closed vessel had to suffocate because, as he described it, they *injure* the air, so that it is no more fit for respiration. He wanted to know whether also plants can affect air. For his experiments he chose small mint twigs, and he observed that these did not *injure* the air, but just the reverse: they *restored* air which was previously contaminated by animals. This observation is contained in a letter which he addressed in 1772 to the president of the Royal Society.

This note was read by the Dutch court physician of the Empress Maria Theresia, JAN INGEN-HOUSZ. He tried to reproduce the experiments but he failed. This caused a long correspondence between PRIESTLEY and him, written in a very polemic style, until both realized that only *green* plants — and green plants only in the light — *restore* air. After this discovery it took several decennia before the underlying process began to become understood. 1842 JULIUS ROBERT MAYER in Heilbronn had formulated the First Principle of Thermodynamics. Three years later he published a small booklet [1]: today we would call it bioenergetics. He did not use the term *energy*; he spoke of *power*. He claimed that power cannot be created *de novo* and that it cannot be destroyed. He realized that there are different forms of power, which can become converted into each other. In his small booklet we find the statement that the green plants catch the power of the sunlight and convert and deposit it in stable form. Today a physicist would choose quite a different formulation, but in fact it is MAYER's idea that electromagnetic energy can be converted into chemical energy.

In 1862 the German botanist JULIUS SACHS in Würzburg observed that illuminated green plant cells contain small granules which he could identify as starch. He assumed that this carbohydrate would be a secondary product, and he concluded that its precursor should be a soluble sugar. SACHS gave the first equation for photosynthesis as it is still used in many of our elementary textbooks:

$$6\ CO_2 + 6\ H_2O \xrightarrow{\text{(light)}} C_6H_{12}O_6 + 6\ O_2 \tag{1}$$

It says that the first product which is formed by illuminated green plant cells is a hexose molecule. Its synthesis starts with 6 CO_2 and 6 water molecules and leads to 1 molecule of a C_6-sugar, which is then polymerized to starch. At the same time we obtain 6 oxygen molecules. This formula cannot be more that a mere balance equation; if it would be correct it would claim that if we put mineral water — that's what we have as substrate — into light, we obtain sugar and oxygen. Nobody would be inclined to postulate this phantastic conversion. Furthermore there is another problem, which was evident even in the 19th century: one of the basic laws in photochemistry says that chemical reactions can only be influenced by radiation, if (at least) one of the reaction partners can absorb the incident light, *i.e.* at least one substrate must be a coloured compound. In photosynthesis both CO_2 and H_2O do not absorb light; all their absorption bands are in the infrared and the ultraviolet region. So we cannot expect any influence of illumination.

In the 19th century already another photochemical reaction had been carefully studied: the photographic process. For this we choose silver halides like silver bromide and silver chloride. The absorption bands of these compounds are in the ultraviolet and the blue region. It was, however, observed that certain dyes, which form a thin film on the crystal surfaces, can *sensitize* these silver halides for longer wavelengths. It is curious enough that one of the first sensitizers which were used by BECQUEREL was chlorophyll extracted from green plants. We may expect that in photosynthesis we have to have a suitable sensitizer which can take up energy and transfer it in the form of excitation energy (see below) either to CO_2 or to the water molecule.

2.   Nature of light

How have we to treat light in photobiological reactions? Since the days of the great Greek philosophers there are two main theories on the nature of light. Both were inseparably mixed with the speculation of the vision, *i.e.* a physiological process. The first one – propagated *e.g.* by PYTHAGORAS – looks like a radar principle. These philosophers thought that our eye is not a passive organ but that it actively emits rays. These were believed to become reflected and returned to our eyes by the objects. PYTHAGORAS and his contemporaries realized that these rays have to have an immense speed: they were well aware that when we open up our eyes during the night we see all the stars immediately. So the big distance between earth and heavens has to be bridged in a fraction of a second. There was at the same time a second hypothesis which we can perhaps describe as a theory of replicas, promoted by DEMOCRITUS, LEUCIPPUS and others. The protagonists of this theory thought that thin pelliculae come off from all objects, pass the space and then hit our eyes which were thought to act merely passively. We may describe the first assumption as a primitive *wave theory*, the second one as the precursor of later *corpuscular* theories.

In the context of a *wave theory* we have to ask: what is the medium in which the postulated waves can propagate? The stoic philosophers already presented the idea that this propagation might be comparable with that of waves on a water surface. It was ARISTOTLE who first discussed the interference of something which we might call an *ether*. Medieval scientists did practically nothing to discriminate between these models, which survived till to the 19th century. Since the work of HUYGHENS the physicists favoured the wave theory. We apply its terminology even today when we use wavelengths, frequencies, amplitudes, etc. There are, however, some phenomena which were not interpretable in the framework of this model. This is *e.g.* the photoelectric effect. Its theory was given (1905) by ALBERT EINSTEIN. To explain it we have to describe light as an ensemble of particles.

If we try to discriminate between those physical phenomena which are better interpreted in a wave model and those which ask for a corpuscular concept, we find that the propagation of light can be best described by waves whereas the emission and absorption of light by matter must be expressed by a corpuscular theory. Since all photobiological phenomena depend on the interaction of light and matter, we should describe them by a *corpuscular hypothesis*. But then we cannot use wavelengths and frequencies any more, which have no place in such a concept. Instead of this we have to speak about energies. In the following text both terms will be used. Fortunately enough it is easy to transform one into the other. We can use EINSTEIN's relation

$$E = h\nu \qquad\qquad (2)$$

So if we have the frequency of light, we have automatically the energy of its photons.

There is another peculiarity which discriminates between pure photochemstry and photobiology. The photochemist studies the behaviour of molecules within a test tube, where all molecules have statistically distributed distances. This is quite different from the situation in photobiology. Here all the sensitizers which we are dealing with as well as all electron donors and acceptors are arranged in surfaces where they possess more of less constant positions. It is therefore often more adequate to use the laws of solid state physics than those of classical photochemistry.

3.   Excited states

What happens when an atom absorbs a photon? To start with the simplest system, the hydrogen atom, we can draw an energy scale (Fig. 1). As zero value we can take the energy of the *ground state*. Then there is another characteristic level: the highest energy which can be stored inside this atom: the *ionisation energy*. We have to supply it (to the ground state) to split the atom into its two constituents: one proton and one electron. Between these two values we find a whole series of energy terms representing possible states of the H atom. Outside these distinguished terms there exists no stable arrangement.

In the case of molecules the situation is more complicated (Fig. 2). Here we do not have just one energy term for the ground state. In all systems with more than one nucleus the constituents can vibrate against each other. Accordingly we observe several vibration terms. From the energetic point of view the distances of these terms are rather small. We can excite vibrations with infrared radiation. In the absorption spectra the vibrational terms cause a *fine structure* which broadens the ground level, and the level of all excited states to a *band*. Under normal physiological conditions, *i.e.* at room temperature, practically all molecules are in their ground state. They are not only in this energy-poorest configuration but also in their lowest vibrational state. Whenever we heat or irradiate them in the infrared region we increase their intramolecular vibrations.

In the term schemes of atoms and molecules we observe a whole series of higher energy states distinguished by different electron energies which are reached by photon absorption. They are characterized by an asteric (*):

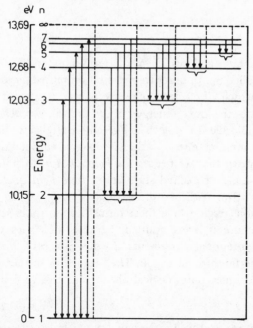

Fig. 1 – Term scheme of hydrogen atoms. The energy difference between the
ground state and the ionization level (ionization energy) is 13.69 eV. Upward
arrows characterize absorption lines, downward arrows emission lines. All transitions ending on the same energy level are characterized as series (after Ref. 67,
modified).

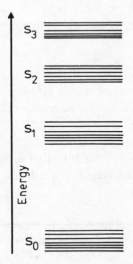

Fig. 2 – Schematic term scheme of a molecule. For the ground state (S_0) and three
excited states (S_1, S_2, S_3) a series of vibration terms (corresponding to infra-red
absorptions) is given; rotational terms are not considered (details see text).

$$A + h\nu \longrightarrow A^* \tag{3}$$

We call them *singlet states* and give them the term numbers S_1, S_2 ... etc, whereas the ground state is called S_0. In order to present a simplified picture we may restrict on arrangements in which all electron spins are paired. If we add enough energy to lift one electron into the next (energy-richer) orbital, we rearrange the charge distribution in the molecule. In molecules like chlorophyll *a* we need less than 2 eV to come from the ground state (S_0) to the first excited singlet state (S_1). In physiological terms we may say that we need a *red light photon* to transform the ground state molecule into the first excited state and a *blue light quantum* to reach the second excited state (S_2). With photons of more energy we can reach even higher levels (in the ultraviolet region). All these terms have a whole series of vibrational levels. Even this picture is oversimplified, because molecules undergo not only vibrational but also rotational movements. So they possess additional levels which can be excited in the far-infrared region. They lead to a further broadening of the absorption bands. For most physiological phenomena this very simple picture may be sufficient.

Important figures are the *lifetimes* of the excited states. If an energy corresponding to a red light quantum is concentrated in one single molecule this represents the same energy concentration as if we would heat this molecule up to 20 000 °C. We cannot expect to keep such a high energy density for a longer time. In fact the lifetime of the higher (energy-rich) excited states is very short, for the S_2 state of chlorophyll *a* it is about $10^{-11} - 10^{-12}$ s, for the S_1 between 10^{-8} and 10^{-9} s. All reactions of excited molecules must occur within this very short time interval.

In *ordinary* chemistry we are dealing with the behaviour of ground state molecules, whereas in photochemistry we are interested in the peculiarities of excited states. For the chemist it is important to realize that the chemical properties of excited states are quite different from those of the ground state. Fig. 3 shows the situation for the first excited singlet state, where the two energy-richest orbitals differ in their energies. If there is a suitable other molecule near enough to accept an electron, the excited molecule acts as a good donor. It is at any case a better donor than the ground state.

Fig. 4 shows that, on the other hand, an excited molecule can also function as an efficient electron acceptor. Whenever an excited singlet state meets in its neighbourhood another molecule with an electron in a sufficiently energy-rich term it can refill its electron *hole* not by the electron which was originally in this position (which is now kept in a higher orbital) but by an electron from the neighbour molecule.

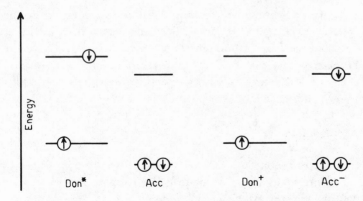

Fig. 3 — Donor properties of an excited singlet state. Transition from the highest occupied term of Don* to the lowest unoccupied level of a neighbour molecule.

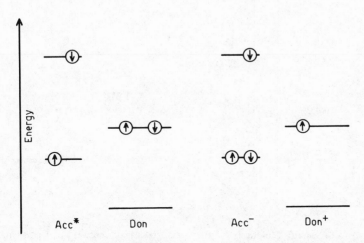

Fig. 4 — Acceptor properties of an excited singlet state. Transition from the highest occupied level of a donor molecule (in its ground state) to the lowest unoccupied level of an excited acceptor (details see text).

4.   Reactions of excited molecules

What happens to an excited molecule? Interesting is especially the reaction of the S_1 state (Fig. 5). By excitation all electrons started from the lowest state of S_0, but they could reach any vibrational term of S_1. After $< 10^{-9}$ s – the life-time of S_1 – the excited molecule reaches its lowest vibrational term again. From this energy level the *lifted* electron falls back to any of the vibrational terms of the S_0 state. The energy difference can be released in the form of heat:

$$A^* \longrightarrow A + heat \tag{4}$$

This would correspond to the situation which we have in the transition from S_2 to S_1. More interesting for many photobiological processes is the release in the form of a photon, a process which we call *fluorescence*:

$$A^* \longrightarrow A + h\nu' \tag{5}$$

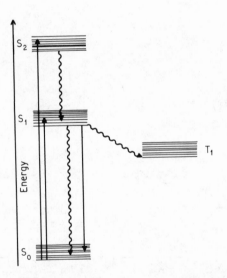

Fig. 5 – Schematic demonstration of possible transitions between different energy states. From ground state (S_0) two absorptions (of energetically different quanta) lead to the excited states S_1 and S_2. From S_2 to S_1 only a radiationless transition (within $< 10^{-12}$ s) can be observed, whereas from S_1 to S_0 either a similar step (wavy line) or the emission of a photon (fluorescence, continuous line) is possible. Furthermore there is a radiationless transition from S_1 to the triplet state T_1 which is characterized by parallel electron spins (see Fig. 6). The term differences of the scheme correspond to those of chlorophyll *a* (details see text).

   The lowest energy gap which we have to bridge in photon absorption presents the transition from the lowest vibrational state of S_0 to the first excited state (S_1). This is, on the other hand, the highest energy which we regain if the electron falls back. Normally it will stop on one of the higher vibrational levels of S_0, *i.e.* the energy which we regain ($h \nu'$) is somewhat smaller than the absorbed energy ($h \nu$). Compared to the absorption peak the wavelength of the emitted fluorescence is shifted to the long wavelength side.

   From this simple picture we understand that an absorption spectrometer measures the *fine structure* of the S_1 state; absorption spectroscopy cannot give any information on the S_0 state. If we want to see the fine structure of the ground state we need a fluorescence spectrometer, because only the fluorescence reflects the wanted details of the S_0 state.

   It can happen that by excitation the spin of the energy-richest electron is reversed (Fig. 6). In this case we obtain a molecule with two parallel electron spins; we call these arrangements *triplet states*. Because molecules never contain electrons with all identical quantum numbers, the PAULI principle forbids that the excited electron returns to its original position. Triplet states have therefore much longer lifetimes than singlets. In the case of chlorophyll *a* the S_1 state has a lifetime between 10^{-8} and 10^{-9} s (see above), whereas the lifetime for the state T_1 is $\sim 10^{-4}$ s. So there is a difference by a factor of 10 000 − 100 000. Many photobiological

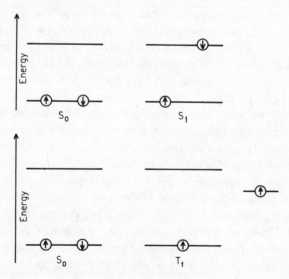

Fig. 6 − Electron spin orientation in singlet and triplet states (details see text).

reactions include a transformation of a singlet to a triplet state. This means always a loss of energy, but, on the other hand, an increased probability for the excited molecule to undergo secondary reactions. In photosynthesis triplet states play no important role. All the energy which is used in thylakoids is the energy of the S_1 state. In the long wavelength — *i.e.* red-absorption-band this means an energy of 1.8 eV, whereas the T_1 state has only ~ 1.4 eV. The $S_1 \rightarrow T_1$ transition would therefore waste a considerable percentage of the principally useful energy of the absorbed quantum.

There are two possibilities to deactivate a triplet state. Increased temperatures can excite higher vibrational states, which become energetically equivalent to the S_1 state. In this case there is a reasonable probability for a renewed spin reversal; the electron can then easily return to the ground state. This means a somewhat *delayed fluorescence*. It may last for microseconds; nevertheless it has the same spectral composition as the spontaneous fluorescence. When this immediate transition into the ground state is *forbidden*, it does not mean that there is no transition at all. There is still a small, but measurable, probability which leads to a *phosphorescence*. Since the bridged energy gap (between T_1 and S_0) is much smaller, the emitted photon contains accordingly less energy. So the light emission by chlorophyll phosphorescence shows a lower frequency (longer wavelength). Furthermore thylakoids always emit some heat.

How much energy do we really need to bridge the gap between the S_0 and the S_1 state? From photochemistry we know that especially systems with conjugated double bonds require rather small excitation energies, so that their absorption bands are in the visible region. Among the most important natural sensitizers are the carotenoids — with a long chain of conjugated $C=C$ double bonds — and macrocyclic compounds like *e.g.* the porphyrins (including the chlorophylls). $C=C$ double bonds are not composed of two identical bonds; they have one σ and one (less stable) π bond. In principle both bond types can be excited. From elementary organic chemistry we know that compounds with a $C=C$ double bond do not show free rotation anymore. This is the reason why for these compounds geometric isomers are found. There are *cis* and *trans* forms. After excitation one of the bonds is split. Each of the two adjacent carbon atoms then bears a surplus electron. During the short lifetime of this state a free rotation becomes possible. In other words: after excitation parts of the molecule begin to rotate, until at the end of its lifetime the double bond is restored. This might happen in the opposite geometric configuration of the molecule, so that a transition from the *cis* to the *trans* form — or *vice versa* — is observed. The reversible transition between *cis* and *trans* forms is well known for the vision process as it works in our retina, but we do not expect exchange of this kind in photosynthesis.

In some cases the absorption spectra of the *cis* and *trans* form differ considerably; *i.e.* the energetic distances between S_1 and S_0 are different for the two isomers. *Cis* and *trans* form can have therefore different colours. If one of the isomers is irradiated the compound can change (and often it does) its colour. If afterwards the molecule is illuminated in its (new) absorption band, it can return (and often it does) to the original colour. In photochemistry this phenomenon is described as *phototropy*. In photobiology there are several reversible light reactions, but in all analyzed cases there seems to be a more complicated conversion (including redox processes).

After the breakage of a C=C double bond it is not necessary, that the original bond is reformed. There may be a rearrangement; so photon absorption can create new molecules.

5.   Energy transfer

All processes mentioned are reactions of the excited molecule itself. On the other hand the energized species must not itself react, but can transfer its excess energy to another molecule. The problem which was mentioned already in the context of the balance equation of photosynthesis (1) is the question whether chlorophyll can absorb a photon and deliver its excess energy either to CO_2 or to water.

In the case of chlorophyll *a* we have a macrocycle with 56 π electrons. This explains why we need a rather low energy to excite this system. In organic solvents it shows two main absorption bands, one in the red, one in the blue region (Fig. 7). They correspond to the two transitions $S_0 \rightarrow S_1$ and $S_0 \rightarrow S_2$ as given in Fig. 5. If we observe the absorption bands of chlorophyll *in vivo* they are much more compli-

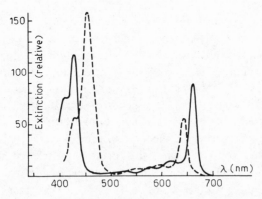

Fig. 7 — Absorption spectra of chlorophylls (in ether). Continuous line: chlorophyll *a*, broken line: chlorophyll *b* (after Ref. 68, modified).

cated. There is no sharp absorption peak, but the band is broadened. With a high dispersion spectrometer we see that the red band region is composed of a whole series of bands [2]. This fact found divergent interpretations. Fig. 8 shows that there may be several chlorophyll *a* molecules bound to complexes of different size. They are connected by water molecules. It is known since years that it is very difficult to isolate water-free chlorophyll. There is still discussion in photosynthesis research whether by this aggregation only double molecules (dimers) or even bigger complexes are formed.

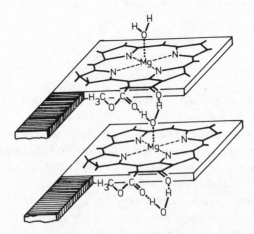

Fig. 8 – Chlorophyll dimer (probably part of a bigger complex). The two porphyrin rings are connected by water molecules. The hatched attachments characterize the phytol chain (after Ref. 69, modified).

The absorption spectrum of chlorophyll *a* (Fig. 7) shows a rather sharp absorption band near 660 nm and another one at about 430 nm; the broken line belongs to the chlorophyll *b* which differs from chlorophyll *a* by just one substituent ($-C=O$ instead of $-CH_3$). In all higher plants it accompanies the chlorophyll *a*. The role of this accessory pigment — which may be absent in some mutants — is still controversial. If we compare this absorption spectrum with an action spectrum of the photosynthesis process we would expect high quantum yields — *i.e.* a low quantum requirement — in the two absorption bands, whereas in the broad intermediate region, the *green gap*, the quantum efficiency should be low. The actually measured data (Fig. 9a) give quite another result: there is a high quantum efficiency in the range of both the $S_0 \rightarrow S_1$ and $S_0 \rightarrow S_2$ transitions but also in the green region. This is one remarkable anomaly in the comparison between action and absorption spectra which asks for a convincing interpretation. There is a second

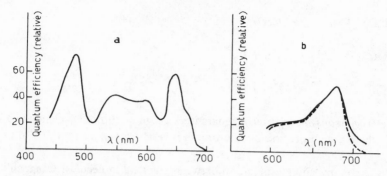

Fig. 9 — Anomalies of the action spectrum of green plant photosynthesis.
a) Photosynthesis rate as function of the incident wavelengths. A comparison with
the absorption spectrum of chlorophyll *a* and *b* (see Fig. 7) demonstrates an un-
expectedly high quantum efficiency in the *green gap,* i.e. the frequency region
which is only slightly absorbed by chlorophyll (after Ref. 70, modified);
b) so-called *red drop* of photosynthesis at the long wavelength end of the chlorophyll absorption band. Details see text (after Ref. 71, modified).

peculiarity in the long wavelength region (Fig. 9b): behind ~ 690 nm where we still
observe a rather good chlorophyll absorption the quantum efficiency is very low, an
effect which is often called the *red drop.*

How can we explain a high quantum efficiency in a frequency region in which
there is practically no chlorophyll absorption? Chlorophyll molecules reemit part of
the absorbed photon energy as fluorescence. In pure organic solvents their fluorescence rate can be rather high, whereas *in vivo* the emission is very weak. Could there
be any efficient reabsorption of emitted photons? The actual yields speak against
this possibility.

Plant cells contain various compounds with conjugated double bonds. There
is practically no photosynthetically active organism which has nothing but chlorophyll *a*; there are always other pigments, which are closely associated with it. In red
and blue-green algae these are molecules closely related to the porphyrins: open
chain tetrapyrrol compounds, the so-called phycobilins. They have an absorption
band between the two bands of chlorophyll *a*. All higher plant cells contain a whole
series of carotenoids; part of them show practically no fluorescence. If they are
irradiated in their absorption band, however, we observe the fluorescence of the
chlorophyll *a*. In contrast to the immediate excitation of chlorophyll *a* there is a
measurable time lag between the excitation of the carotenoids and the reemission of
energy from chlorophyll [3]. So we may expect an energy transfer from the excited
carotenoid to the chlorophyll and then a reemission from this secondarily excited
molecule:

$$Car \quad + \quad h\nu \quad \rightarrow \quad Car^* \tag{6}$$

$$Car^* \quad + \quad Chl \quad \rightarrow \quad Car + Chl^* \tag{7}$$

$$Chl^* \qquad \qquad \rightarrow \quad Chl + h\nu' \tag{8}$$

This *heterogeneous energy transfer* guarantees a high quantum efficiency in a spectral region, where the absorption of chlorophyll is very low.

Is there any energy migration between *identical* molecules? Can a chlorophyll molecule transfer the absorbed energy to an identical neighbour according to

$$Chl_1^* \quad + \quad Chl_2 \quad \rightarrow \quad Chl_1 \quad + \quad Chl_2^* \tag{9}$$

The problem is how to measure this type of a *homogeneous* energy transfer. If we illuminate a cuvette with a very dilute chlorophyll solution we have to realize that the lifetime of the excited pigment is very short. Between the absorption and the reemission of a photon only a few nanoseconds will pass. Within this short time interval the orientation of the molecules will not severely change. If we irradiate the dilute solution with linear-polarized light we should therefore expect that the emitted fluorescence is also linear-polarized. The actual experiment confirms this prediction. If the concentration of the chlorophyll is stepwise increased there is a critical density at which the polarization of the reemitted light disappears. The sudden change from polarized to unpolarized emission demonstrates a very strong dependence of the transition probability on the intermolecular distance. Studying this phenomenon, FÖRSTER [4] could demonstrate, that the probability (p) for the exchange of excitation energy between neighbour molecules is inversely proportional to the 6th power of the distance (r)

$$p = \text{const} \, (1/r^6) \tag{10}$$

So the smaller the distance of the chlorophyll molecules, the higher the probability of excitation transition. If we calculate which distances can be expected for such a migration [5] we obtain values in the order of nanometers (till to $\sim$ 8 nm). Actually the distances of the chlorophyll molecules within the thylakoid surface (see below) are definitely smaller. So there is a very high probability of energy transition from Chl* to unexcited neighbour molecules. We may call this special case in comparison to mechanical analogues a *resonance transfer*. This experiment demonstrates an energy migration from one sensitizer molecule to another. If the molecules are not identical, the only precondition for an efficient transfer is that the secondary sensitizer needs for its excitation less energy than the first one.

It may be asked whether there is any evidence that models of this kind may play a role in biology. It is worthwhile to mention a meanwhile famous experiment which was done in 1932 already by EMERSON. At that time the technique of short light flashes became available. EMERSON applied strong light flashes exciting every chlorophyll molecule in the chloroplast to determine how many oxygen atoms per chlorophyll molecule were released. It turned out that there must be ∼ 2000 chlorophyll molecules to one oxygen molecule [6]. This caused EMERSON to postulate a *chlorophyll unit* in which a whole ensemble of chlorophylls cooperates. All photons which are absorbed within this area have to be directed to some special place where they induce the evolution of molecular oxygen. It could well be that this migration is a pure resonance transfer. On the other hand there are some more conceivable models.

The necessary time for the excitation energy to hop from one excited molecule to another by resonance is ∼ 10^{-12} s. There is another mechanism which has to be discussed in photobiology: if the term schemes of two molecules are practically identical there is a bidirectional transfer (*excitation migration*) which takes much less than 10^{-12} s (Fig. 10). At any case a chlorophyll molecule in the state S_1 can transfer the excitation energy to many other molecules before the ensemble loses its excess energy. So an area of several hundred sensitizer molecules would be just a conceivable unit. Fig. 11 gives a textbook model. It combines two ideas; first: that there is a special chlorophyll *a* molecule to which one electron donor and one electron acceptor are attached, whereas all other sensitizer molecules − *i.e.* the rest of the *chlorophyll unit* molecules − is only engaged in energy absorption and transfer; second: that the unit is heterogeneous and contains both chlorophylls and carotenoid molecules. For lower plants we would have to modify this scheme and to replace carotenoids by phycoerythrin and phycocyanin molecules.

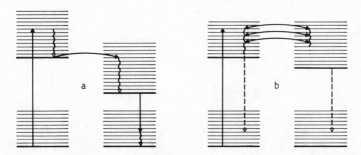

Fig. 10 − Unidirectional (a) and bidirectional (b) excitation transfer between neighbour molecules. (a) characterizes the resonance transfer (FÖRSTER mechanism) which requires ∼ 10^{-12} s, (b) the faster (< 10^{-12} s) exciton migration (details see text; after Ref. 72, modified).

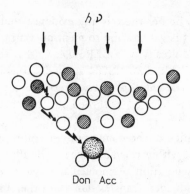

Fig. 11 — Model for a pigment antenna of the chloroplast. The antenna consists of chlorophylls (open circles) and accessory pigments (carotenoids or phycobilins; hatched circles). Both transfer the absorbed energy to a reaction center with a specially bound chlorophyll *a* molecule (dotted circle) attached to both a donor and an acceptor (after Ref. 73, modified).

For some time only this model had been discussed in photobiology. We have to consider quite a different possibility, however. If we increase the concentration of certain dye solutions like methylene blue, we observe a change of the absorption spectrum. We have to recognize that BEER's law becomes invalid. This effect is caused by the formation of double molecules above a critical concentration level. These dimers have a different absorption spectrum. In some cases dimers are only formed if one of the molecules is excited. This means: in the S_0 state there are only single molecules. But if there is a $S_0 \rightarrow S_1$ transformation we obtain double molecules which we call *excimers*. Whether we have such excimers in photosynthesis is an open question.

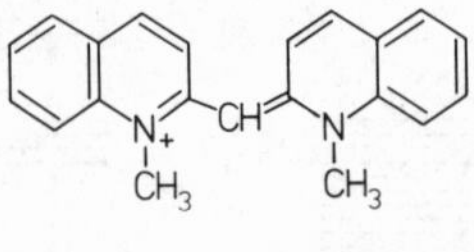

Fig. 12 — Pseudoisocyanin.

Fig. 12 gives the structural formula for a very special dye, the pseudoisocyanin. At first sight it has nothing to do with chlorophyll. It possesses a system of conjugated double bonds; like the porphyrin head of the chlorophyll, pseudoisocyanin is planar. So we may regard it, in an oversimplified manner, just as a flat plate. If its concentration is increased its solutions attain a high viscosity. The aggregation does not stop at the level of double molecules but leads to much bigger complexes. In a concentrated solution these may form long chains with ~ 1 million molecules. We call them *reversible polymers*; if we dilute their solution we regain the monomers. The optical properties of these polymers are very remarkable: they show a pronounced resonance fluorescence, *i.e.* the emitted light has the same frequency as the absorbed radiation.

A well-known effect in photochemistry is the quenching of fluorescence. There are many compounds which prevent any reemission of photons by fluorescing molecules. Normally one quencher molecule to one fluorescing molecule is required. In the case of pseudoisocyanin SCHEIBE observed that we need only a single quencher molecule to the whole dye chain, *i.e.* to 10^6 monomers [7]. Effective quencher molecules are *e.g.* different phenols. Brenzcatechol added to the pseudo-isocyanin polymer will be bound at any place of this long chain. If we observe that afterwards there is no reemission of energy, there must be an energy migration. We have to imagine that the transfer works without any loss. So we may call this aggregate an *antenna*. This is comparable to the antenna in radio receivers, where electromagnetic energy is collected. In the case of photosynthesis electromagnetic energy in the spectral region of visible light is harvested and directed to a distinguished point which we may call a *reaction center*. We can put up two-dimensional layers of this special dye by dipping glass into its concentrated solutions. So it is not necessary to discuss a uni-dimensional model only. Compared to the resonance transfer this would be quite another model. To summarize the essence: there is an antenna − a structure which collects electromagnetic energy − and a reaction center where secondary reactions preventing the reemission of energy happen.

Finally there is a third model, which we know from oscilloscope or television screens. Going from atoms to molecules we realized a higher number of terms: first vibrational terms, then in addition rotational terms. In a crystal lattice the density of the terms is so tight, that we cannot resolve them with normal spectrometers. They practically merge into a band which we call the *valence band* (Fig. 13). In irradiated crystals like the zinc sulfide of a television screen, single electrons reach a higher energy level, which we call the *conduction band*; in between the two levels there is a big gap. Whenever the available energy is sufficient to bridge it, one electron moves into the originally empty conduction band. Within an electric field the irradiated crystals show an electric current (*photoconductivity*). In the dark the same salt behaves as in insulator.

The term scheme can be more complex: if we compare different oscilloscope or television screens we see that all consist of a zinc sulfide lattice, but that the colour of the emitted light differs considerably. No commercially produced screen is pure zinc sulfide. Deliberately different atoms, *e.g.* copper, are included. These *contaminations* have quite different energy terms. If we irradiate such a *doped*

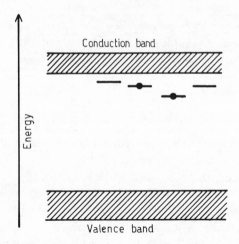

Fig. 13 — Bond structure of a (doped) semiconductor. Between the valence band and the energy-richer conduction band a few (partly filled) electron traps (details see text).

lattice, electrons from the conduction band fall down and stay in the additional *traps*. We may compare this situation with the transition into the long-living triplet state.

The lifetime of electrons in the conduction band is much longer than that of excited states in isolated molecules; it can reach the order of minutes. There are two possibilities for the return of the conduction band electrons into their *holes*. Heating the lattice will supply the missing energy to return into the conduction band. From there they reach the valence band. We may compare this with the return of the electron from a triplet state back to the S_1 state and finally to the ground state. The other possibility is the immediate return to the hole which is linked to a smaller energy release. There should be an interesting consequence: If we introduce different contaminations like copper, thallium or other metal ions, we have traps with different energy differences to the conduction band. Warming up a doped semiconductor which had been irradiated at low temperature we will supply the necessary thermal energy to lift the trapped electron to the conduction band at a characteristic temperature. It then falls back emitting a photon. Before this happens for a second electron in a deeper trap, we need a somewhat higher energy. If we draw the intensity of the emitted light as a function of the lattice temperature we observe characteristic *glow curves* (Fig. 14).

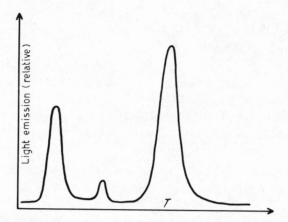

Fig. 14 – Schematic sketch of a *glow curve* as given by doped semiconductors (details see text).

We are inclined to ask: have we to discuss this possibility in photobiology? Is there any indication that in biological systems might be anything like a semiconductor? The small number of sensitizers in a particle may seem too small for this effect. If we isolate light-sensitive particles, like chloroplasts or retina cells, we can perform a very simple experiment: we take an insulator like a sapphire crystal or a block of paraffin as support for a dry film of the biological material. When we connect its ends to metallic wires, an amperemeter and a battery (Fig. 15) we measure (in the dark) a very high ohmic resistance. In the light, however,

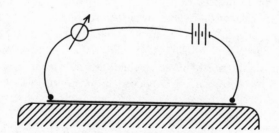

Fig. 15 – Experimental set-up for the demonstration of photoconductivity in dried biological material (details see text).

we observe significant electric currents, *i.e.* we obtain photoconductivity [8]. If we cool a chloroplast suspension on a supporting material to the temperature of liquid nitrogen, irradiate it and then warm it up, we observe a type of glow curve like with a semiconductor crystal. So there are two phenomena which could be explained by a *semiconductor model*. It means we cannot exclude that even in biological systems we have to apply this model. It remains completely open, whether reversible polymers, semiconductors or resonance migration would be the correct picture. At any case we may say, that the migration works with a very high efficiency.

6.   Chloroplast structure

It is, of course, tempting to look for a morphological equivalent of the antenna. If it contains hundreds or even thousands of chlorophyll molecules we should see it under the electron microscope. What is the structure of a chloroplast? Since more than 100 years it was observed that the sensitizers of the plant cells are not homogeneously distributed within the cell but that they are concentrated in special organells (*chloroplasts*), which we can see under the light microscope. These green particles are tiny units with a diameter between 5 and 7 μm, *i.e.* practically the same diameter as our red blood cells. They include all the sensitizers, *i.e.* both chlorophylls and carotenoids. For a structure of this size we have to realize that the light microscope resolves structures down to $0.2 - 0.3$ μm. So if there would be a substructure of this dimension we should see it.

Chloroplasts squeezed from leaves show, that the sensitizers are not evenly distributed within these organells. This was observed many years ago already; all the pigments seem to be concentrated in so-called *grana*, which are dispersed within a colourless *stroma*. It soon became evident, that chloroplasts possess a very high concentration of lipids. Since we know that both chlorophylls and carotenoids are easily lipid-soluble, it seemed conceivable that the green and yellow pigments are dissolved in small oil droplets and that the rest of the plastid is just a protein matrix. We cannot expect more information from studies with the light microscope. When the electron microscope with its much higher resolution power was applied to biological research, one of the first photographs taken was that of an isolated chloroplast of a tobacco leaf. This photograph did not give much more information; it showed, however, that the believed droplets did not exist. Grana seemed to be rather rigid structures. It remained, however, quite impossible to say more about these structural units.

In the meantime the biologists learned how to include both intact cells and isolated organells into polymers like plexiglass. They have to be suspended in monomer solutions, which are afterwards polymerized to glassy substances. Together with this supporting material they can be cut to very thin slices which are transparent to electron beams. With a suspension of chloroplasts we realize that there are no droplets and that the so-called grana are complicated layered structures [9]. These must not be compared to a staple of paper. With high magnification we see that there are no open rims, but that the structural subunits are compressed vesicles. They received a special name: *thylakoids*, which means *sac-like structures*.

For a biochemist this observation is very important. Compressed vesicles contain an internal space which might be chemically different from the external medium. We can combine the pure electron microscopical observation with enzymatic treatments of the preparations. By this means it turned out, that the dark lines are protein-rich, and that the material between these lines must be rich in lipids. Thylakoid membranes are not structure-less. Taken perpendicularly to the thylakoid surface, micrographs do not show a smooth lipid film but a *cobblestone-like* structure [10].

The smallest subunits, the *cobble-stones* themselves, can be isolated *e.g.* after ultrasonic vibration. These particles were once thought to represent the smallest units which can perform the light-induced electron transport. So they were called *quantasoms*. Today we are convinced that this assumption is not correct and that the particles are no part of the antennae. If we measure that there are several hundred chlorophyll molecules per reaction center we have to regard this figure as a purely statistical value.

Fig. 16 tries to incorporate all observations into a scheme: We are convinced that the chloroplast is surrounded by a double membrane. In its interior the thylakoids form staples, which we see as grana. There are also thin lamellae which bridge the gaps across the stroma (stroma thylakoids). It is still a matter of discussion, whether thylakoids are really isolated vesicles or whether their internal spaces are interconnected. Fig. 17 contains the approximate dimensions for a spinach thylakoid. Its diameter may be given with 500 nm; the inner space is very thin, so that it seems doubtful, whether there is really *free* water in between. Within the membrane we have ~ 100 000 chlorophyll molecules.

Apparently there are certain *gradients* across thylakoid membranes. Thylakoids are so small that it is not appropriate to use thermodynamic terms like a pH value. If we would calculate the probability to detect a free H_3O^+ ion inside the inner space we would find it < 1. So a pH value is nothing but a statistical figure.

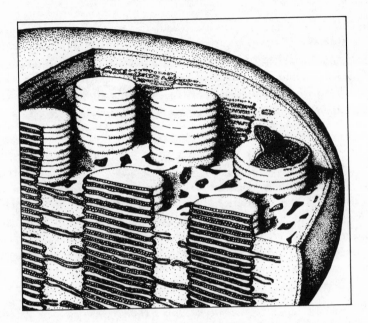

Fig. 16 — Schematic view of a chloroplast. The drawing demonstrates the surrounding chloroplast (double) membrane and thylakoid staples (grana) which are interconnected by extended compressed vesicles (stroma thylakoids). The thylakoid surfaces are covered with tiny particles (details see text; after Ref. 74).

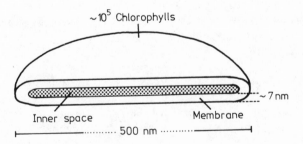

Fig. 17 — Dimensions of a thylakoid (after Ref. 75, modified).

7. Primary charge separation

What happens when the electron energy absorbed in the antenna reaches the reaction center? At this distinguished point we obtain an electron exchange, *i.e.* a first redox reaction. We may expect that the primary reaction is an *electron exchange* between a donor and an acceptor molecule.

$$\text{Don} + \text{Acc} \xrightarrow{h\nu} \text{Don}^+ + \text{Acc}^- \tag{11}$$

For the chemist the question remains, which compounds function as partners. This is, indeed, very difficult to decide.

Starting with the term scheme for donor and acceptor we can ask for the necessary energy to separate one electron from the donor. As to be seen from Fig. 18 this is the *ionization energy (I)*. Theoretically this ionization primarily leads to an oxidized donor and a free electron:

$$\text{Don}^* \longrightarrow \text{Don}^+ + e^- \tag{12}$$

The other decisive value is the *electron affinity* (E_A) of the acceptor. If this is bigger than the ionization energy of the donor, we gain energy if the electron is

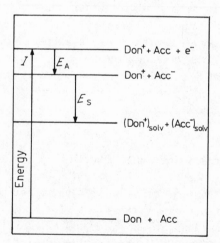

Fig. 18 — Term scheme for an electron exchange between a donor and an acceptor molecule. *I*: Ionization energy, E_A: electron affinity, E_S: solvation (in the case of aqueous system: hydration) energy. Details see text (after Ref. 76, modified).

exchanged. In endergonic photoprocesses (like in Fig. 18) the energy difference between the ionization energy of the donor (in its ground state) and the electron affinity of the acceptor is negative. Reactions of this kind require the previous excitation (see below). At any case we end up with a pair of an oxidized donor and a reduced acceptor:

$$Acc + e^- \longrightarrow Acc^- \tag{13}$$

At first sight one could assume, that if both the ionization energy and the electron affinities are known, we can calculate the released (or stored) energy. This is, however, incorrect. By the electron transfer two radicals, *i.e.* two charged species, are produced; both will become hydrated [11]. This releases hydration (solvation) energy (E_S) both for the donor and the acceptor. Unfortunately these values are practically unknown for biological systems — at least *in vivo*.

Molecules in their ground state have a high ionization energy but a much smaller electron affinity. In the excited state one electron attains a higher energy level. This corresponds either to the energy difference ($S_1 - S_0$) or ($T_1 - S_0$). It is obvious that to ionize an excited molecule we need less energy than to ionize the same molecule from its ground state. On the other hand the electron affinity is much higher, because the possible energy gain is given by the span between the ionization level and the ground state. So excitation changes the relation of ionization energy to electron affinity.

With irradiated sensitizers we can in principle expect two different reaction sequences: we can discriminate between *exergonic* and *endergonic* processes (Fig. 19). In a case like vision we have an exergonic process, *i.e.* the supplied energy must only be sufficient to overcome an activation energy barrier. In the case of photosynthesis — and other light-induced reduction processes — we have the conversion of electromagnetic energy into the chemical energy of bonds. This is a typical endergonic process.

8.   Secondary reactions after charge separation

The simplest living system in which a photo-induced transformation takes place is represented by some bacteria [12]. These autotrophic microorganisms contain sensitizers very similar to the chlorophyll of green plant cells. In comparison to real chlorophyll the so-called *bacteriochlorophyll* has a somewhat smaller number of π electrons. Here the ring of conjugated double bonds is interrupted; furthermore there are some other substituents (Fig. 20). This shifts the absorption spectrum of the bacteriochlorophyll more to the infrared region. The energy content of

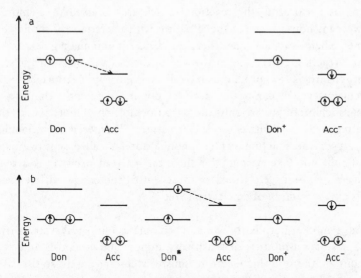

Fig. 19 — Electron exchange between donor and acceptor molecules. a) Exergonic reaction: highest occupied level of the donor higher than the lowest unoccupied term of the acceptor; b) endergonic reaction: highest occupied level of the donor lower than the lowest unoccupied term of the acceptor. Here the supply of excitation energy is required (after Ref. 77, modified).

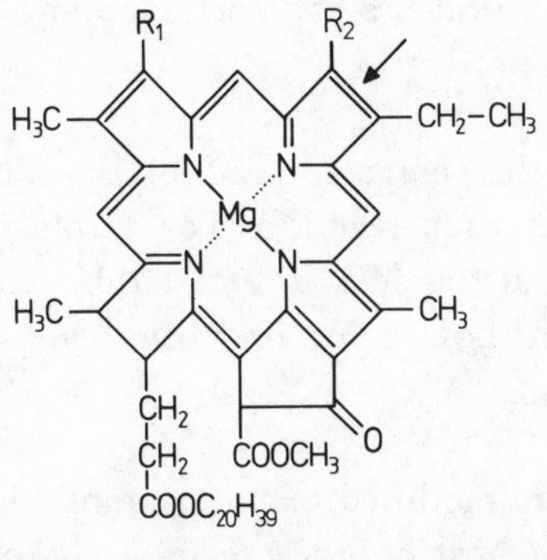

Fig. 20 — Structural formula of chlorophylls. Chlorophyll a: $R_1 = -CH=CH_2$, $R_2 = -CH_3$; chlorophyll b: $R_1 = -CH=CH_2$, $R_2 = -CHO$; bacteriochlorophyll a: $R_1 = -CO-CH_3$, $R_2 = -CH_3$. In this molecule the $C=C$ double bond marked by an arrow is reduced to a single bond.

photons which can be absorbed by bacteria is accordingly lower. There are less than the 1.8 eV, which lead to the excitation of chlorophyll a. With this smaller energy bacteria cannot perform all the reactions which are observed in a chloroplast. In principle we can, however, treat the situation with bacteria very similar to that in higher plants. There is an antenna, there are different antenna pigments, the energy migrates to a reaction center, and at this center there is a primary charge separation. As in green cells irradiation produces an oxidized donor and a reduced acceptor. The problem remains, which compound acts as donor. Apparently this is a specially bound bacteriochlorophyll. So only the bacteriochlorophyll molecules in the antenna may be regarded as sensitizers in a strict sense. The special bacteriochlorophyll, which is part of the reaction center, should not be called sensitizer, because a sensitizer should not take part in the chemical reaction itself. It is a real donor, which delivers one of its electrons to an acceptor; the energy difference between these two is only a fraction of an electron volt.

There is another peculiarity which discriminates biological systems from a test tube: if we irradiate a mixture of sensitizers, donors and acceptors in pure solution the directions of the resulting electron transfer steps are purely statistical; there is never a preferred direction. Quite the contrary in biological systems: here we face a *vectorial* transport.

Many biological donors and acceptors are separated by a *membrane*. With the transfer of electrons from the donor to the acceptor site we build up an *electrostatic field*, which represents an energy. Since this field energy has to be taken from the excitation energy, only the rest of the energy can perform a redox reaction. So we cannot expect that the difference of the standard redox potentials between a donor and an acceptor system can bridge a gap corresponding to the excitation energy: it must be lower.

What happens after the charge separation? In principle there are two possibilities: there can be a direct back reaction. The surplus electron can return to the oxidized donor; this is what happens in a test tube with very high probability. It is really a serious question, how living organisms can, at least to a certain extent, prevent this type of reaction.

For bacteria there are modified back reactions which lead to ATP formation – and the loss of energy as heat or in the form of photons (*delayed light emission*). The formation of ATP makes the charge transfer a cyclic one.

The other possible way for the surplus electron is to reduce coenzymes like NAD^+. If this occurs, there remains the serious question, what happens to the electron hole of the donor. In all bacterial systems it has to become refilled by substances which the organism must find in the medium. In many bacteria these are H_2S or SH^- anions. These simple donors may, however, be replaced by organic compounds like alcohols or aldehydes. This may be taken as a peculiarity of the simple bacterial systems.

Higher plants are independent from external electron donors. They have an internal system of redox couples which is replenished by light-induced reactions. By this means the surplus electron can be used with a much higher efficiency to reduce fixed CO_2. If we cultivate algae or higher plants in complete darkness, only a few of them, like gymnosperms, produce chlorophyll. After a short irradiation even these plants, which look green, cannot perform photosynthesis. They need a certain period of illumination to develop the ability for the necessary electron exchange steps [13]. Chloroplasts irradiated for half an hour work with only one photosystem. The absorption spectrum demonstrates, that we need 1.8 eV to bridge the gap between S_0 and S_1. The primary donor is apparently chlorophyll a, but — very similar to the bacterial system — it is a specially bound chlorophyll. By this bonding the absorption spectrum of the sensitizer is shifted to the long wavelength side; the actual absorption maximum is near 700 nm, whereas in a pure solution this is near 660 nm. We call the bound chlorophyll a P 700. Like the bacteriochlorophyll of the bacterial reaction center, P 700 is no sensitizer; it is a real electron donor. The first acceptor (X) seems to be a bound *ferredoxin*, a compound with iron bound — unlike in porphyrins — to sulfur atoms. Ferredoxins are well-known for many organisms [14]; there is good evidence that there are both soluble and bound forms. If bound ferredoxin is placed on an energy scale, its standard potential is given with $\sim - 600$ mV.

It must be discussed, whether it is appropriate to describe the primary photochemical processes in terms of the redox state. The *redox potential* is a thermodynamic value; the standard potential is taken for a ratio of the oxidized to the reduced form of 1:1. At a single reaction center the donor is either oxidized or reduced. The same happens with the acceptor. The potentials which we find listed in our encyclopediae and textbooks for a 1:1 mixture are valid for *dissolved* molecules and ions only. We know that *e.g.* the redox potentials of several cytochromes, especially cytochrome c, just by binding to a high-molecular compound are shifted for values up to 1 V. In the case of thylakoids we face another problem: if we regard the thermodynamic value of the redox potential we have to realize the lack of constancy. If we start with a certain gap as given by completely oxidized ferredoxin and completely reduced P 700, then during the light-induced charge exchange part of the P 700 becomes oxidized, part of the acceptor becomes reduced. The percentage of the reduction is a function of the intensity of the incident light. This means: the redox potential difference becomes a function of the light intensity. This makes all energetic discussions very complicated. We better look for some more reliable values. These are the *ionization energies* and the *electron affinities*. What we have to know are the ionization potential of the donor and the electron affinity of the acceptor. One difficulty already mentioned (see above) is the necessity to consider the *solvation energies* of the produced radicals. Furthermore we do not have the relevant values for the most important redox systems. This may be taken as the reason why we are still working with the redox potentials.

Keeping all these restrictions in mind we may say that P 700 (in organic solvents) has a potential between + 0.4 and + 0.5 V [15], whereas artificial acceptors which can replace the bound ferredoxin can have redox potentials up to − 600 mV. By the light-induced electron transfer we therefore convert ∼ 1 eV. Where is the rest (∼ 0.8 eV) of the excitation energy? Part of it must be localized in the field which is created by the charge transfer across the membrane. The primary electron acceptor of the photosystem is not exactly known. It transfers its surplus electron to free ferredoxin. For some time biochemists assumed that the ferredoxin itself would be the primary acceptor, but after switching off the light there is still a reduction of ferredoxin. This could, of course, not happen if this redox system would be the primary acceptor. So there must be (at least) one precursor of ferredoxin. Ferredoxins are characterized by their very negative redox potentials. They can reduce both flavoproteins and pyridinnucleotides. Besides this there is, as in bacteria, a possibility for a cyclic pathway, *i.e.* electrons can be rechanneled to the oxidized P 700 (Fig. 21). For this pathway we need a special cytochrome, the cytochrome b_6 which reacts with a special quinone, the *plastoquinone*. That this often is written as a *plastoquinone pool* reflects the fact that there are at least 20 plastoquinone molecules per P 700 [16]. The redox energy difference between these compounds is

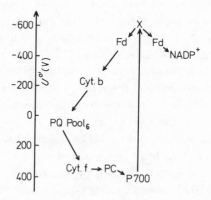

Fig. 21 — Electron transfer in photosystem I (after Ref. 78, modified). X: Primary electron acceptor, Fd: ferredoxin, PQ: plastoquinone, PC: plastocyanine. All other redox couples as given in the text.

big enough to explain the phosphorylation of ADP. So this system has the possibility either to produce ATP in a cyclic electron pathway (*cyclic photophosphorylation*) or to use electrons for the reduction of CO_2.

What happens to the electron hole at P 700$^+$; how can it become refilled? In bacterial systems we need an external electron donor (see above); this is unnecessary for green plant cells or isolated chloroplasts. Inside plastids we have a system of redox couples which we may write as Q_1, Q_2, Q_3. These are three *quinones*; it is not decided whether all are plastoquinones (*i.e.* chemically identical) and whether all have the same effective redox potential (apparently near zero). But besides these quinones there are two metal-containing redox systems: one is a cytochrome (cytochrome *f*), the other one a copper-containing compound, the *plastocyanine*. Both have a nearly identical standard potential near + 0.37 V [17]. From the two last compounds the electron goes to the oxidized P 700.

In a greening chloroplast this reaction can run as long as there are electrons available in the pool of the different quinones, in the cytochrome *f* and the plastocyanin. This is, however, a very small electron reservoir, *i.e.* after a very short time the reaction will stop. This is the situation which we observe immediately after the beginning of the greening process. If we wait for some hours we can see that the electron transport is running with a much higher turnover rate.

9.   CO$_2$ incorporation and reduction

The first formula as given by JULIUS SACHS (see above) demonstrates the reduction of CO_2. There are 6 CO_2 and 6 H_2O which produce a primary C_6 compound: a hexose $C_6H_{12}O_6$. Why not divide the whole sequence by 6; then 1 CO_2 and 1 H_2O would give 1 CH_2O. This compound is the *formaldehyde*. More than 100 years ago the Russian chemist BUTLEROV could demonstrate that under irradiation formaldehyde undergoes a polymerization; this reaction leads to a mixture of sugars (*formose*). The first hypothesis of the chemical pathway of photosynthesis was, indeed, a formaldehyde hypothesis as given by BAEYER in 1870 already [18]. The biologists were well aware that formaldehyde is very poisonous. So they had to assume, that the pool of the CH_2O always remains very small. Nevertheless this hypothesis survived till to the forties of this century. At that time it became possible to study the chemical pathway of photosynthesis with radioactive carbon.

Unfortunately the first radioactive carbon isotope which became available at that time was the ^{11}C. It has a mean half-life of $\sim$ 20 min. In these years there were no fast separation procedures for the fractionation of complicated mixtures available. Before the invention of paper chromatography there were only columns on which the fractionation procedure could be tried. This took such a long time, that the only result of these early studies was the realization that it is not the CO_2 itself which is reduced but a carboxyl group. The first step is the incorporation of carbon dioxide into a carboxyl group [19]:

$$R-H + CO_2 \longrightarrow R-COOH \tag{14}$$

It was difficult to identify the natural CO_2 acceptor R—H, which may be written as a residue with hydrogen. The formation of the first photosynthesis product is a rather fast reaction. It was possible to cut down the fixation time till to half a second. Even within this short time interval radioactive carbon is incorporated into a COOH group. We cannot call this *photosynthesis*. The CO_2 incorporation is a dark reaction which is by no means restricted to plant cells, but which even in animal tissues can be observed. The decisive step in photosynthesis is the reduction of this carboxyl group:

$$R-COOH + XH_2 \longrightarrow R-CHO + X + H_2O \tag{15}$$

We have to postulate a strongly reducing agent which can be a ferredoxin or a reduced pyridinenucleotide; the resulting reduction product is R—CHO. Furthermore we get water and regain the oxidized electron (or hydrogen) donor X.

With the short-living isotope ^{11}C it was impossible to study this pathway in more detail. Fortunately enough a few years later the ^{14}C with a half-life time of 5600 years became available. This made it possible to identify the primary fixation product R—COOH. The first result was, that with a fixation time of less than 1 second the ^{14}C is more or less concentrated in *one* radioactive substance. This primary compound is the *3-phosphoglyceric acid* (Fig. 22). The ^{14}C is incorporated

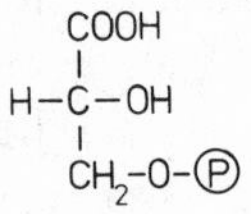

$$
\begin{array}{c}
\text{COOH} \\
|\\
\text{H--C--OH} \\
|\\
\text{CH}_2\text{--O--}\textcircled{P}
\end{array}
$$

Fig. 22 — 3-Phosphoglyceric acid (first stable CO_2 fixation product).

into its carboxyl group. This observation misled the biochemical research for several years: if the first fixation product contains 3 carbon atoms, one would expect an acceptor with 2 C atoms. This, however, does not exist. It turned out that the first fixation product is a very unstable β-keto acid. It cannot be isolated from the cell; in aqueous solution it is decomposed. Nevertheless it was possible to show that the precursor is a five-carbon sugar, the double-phosphorylated form of a pentose (ribulose-bisphosphate) (Fig. 23). So we face a curious situation: to form sugar we must already have a carbohydrate. This pentose molecule exists in two tautomeric forms. The reactive species is the so-called enediol form with a C=C double bond between the carbon atoms 2 and 3. Biochemists know that it is easy to add CO_2 on double bonds of this type. By this we end up with a branched-chain compound as given in Fig. 24, which breaks between the carbon atoms 2 and 3. The result are two C_3 compounds: the *phosphoglyceric acid*. This C_3 compound is reduced to a three-carbon sugar: glyceraldehyde-phosphate. We see that reaction (1) given at the beginning may just be written for historical reason. The first photosynthesis product is not a hexose but a triose molecule. It forms an equilibrium system with dihydroxyacetone-phosphate. The formation of the first stable C_6-sugar is a secondary step.

Since we need already a sugar with five carbon atoms as CO_2 acceptor we always have to rebuild this compound. This happens in a series of C_2-and C_3-transfer reactions, the so-called CALVIN *cycle* [19]. We can write it as a very simple scheme (Fig. 25). There are $6 C_5$-molecules which add $6 CO_2$; by this means we obtain

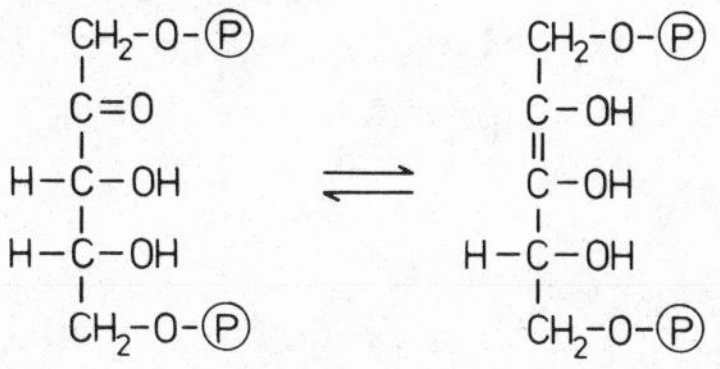

Fig. 23 — Ribulosebisphosphate (equilibrium between normal, keto and reactive enediol form).

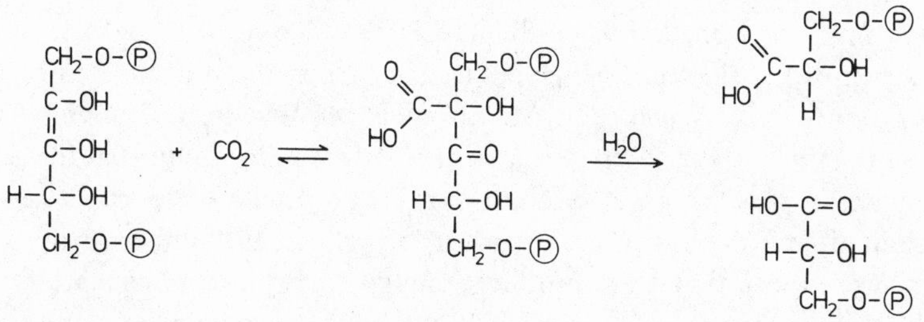

Fig. 24 — CO_2 fixation to the primary acceptor (enediol form of ribulosebisphosphate) and decomposition of the unstable fixation product (branched-chain β-keto-acid) into two molecules of 3-phosphoglyceric acid.

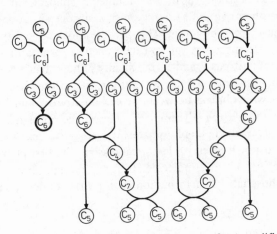

Fig. 25 — CALVIN cycle (details see text; after Ref. 76, modified).

unstable keto-acids which decompose to the C_3-acid from which the first photo-synthetic product is formed (see above). Only two of the twelve resulting triose molecules combine to a storeable product. The rest is recycled and gives the C_5-acceptor again.

Strictly speaking the CALVIN cycle has nothing to do with *real* photo-synthesis. All steps are dark reactions. The important compound which is produced in photosynthesis is the strongly reducing agent XH_2 which we need to reduce a carboxyl group to the carbonyl group of an aldehyde. This XH_2 is produced by a light-induced electron transfer.

An energetic calculation easily reveals that we need one other compound: ATP. So part of the electrons which are transferred to acceptor molecules must flow back to produce ATP. When the CALVIN cycle was first formulated biologists were convinced that it is the mitochondria of the cells which produce all the ATP. If we observe plant tissues we have to admit that cells which are especially active in the production of carbohydrates, possess a surprisingly small number of mitochondria. How can we expect that they produce enough energy-rich phosphate to keep the CALVIN cycle running? It was ARNON who could demonstrate that in green plant cells the same electron pathway as in photosynthetic bacteria operates [20]. There is a back reaction between a reduced acceptor and an oxidized donor which forms ATP (*cyclic phosphorylation*).

10.   Quantum requirement of photosynthesis

The quantum yield of photosynthesis has been discussed for many, many years [21]. How many photons do we really need to release one oxygen or to reduce one CO_2 molecule? In most laboratories the values were converging against ~ 8. WARBURG, however, claimed that we need only *one* photon for the process [22]. If we repeat his experiments they seem to be correct *within the first seconds*. At the beginning of the light period we really face a surprisingly high quantum efficiency. After a few seconds it decreases to $\sim 1/8$. This is a very unusual situation; in photochemistry we know of no reaction with a variable quantum efficiency. So we have to postulate that there are in fact *two* different photoreactions: one which can only be observed at the beginning of the light period (with a quantum requirement equal to 1). This process does apparently not contribute to the normal photosynthetic electron transfer, which requires 8 photons.

The two light reactions of photosynthesis

Discussing the action spectrum of photosynthesis we realized two anomalies: one — the high quantum yield in the *green gap* — has already been discussed (see above); there remains the remarkable *red drop*: an unproportional decrease of the quantum yield in the long wavelength region (compare Fig. 7 chlorophyll *b*). It was first observed by EMERSON, that light of ~ 700 nm is used with a very low quantum efficiency. If very small intensities of shorter wavelength light (in the region of 660 - 680 nm) are added, the photosynthesis rate increases significantly more than to be expected by the simultaneous action of the two absorbed energy bands [23]. We now call this phenomenon the EMERSON *effect*. EMERSON explained it by the assumption of *two* photoreactions. He postulated one light-induced process as described above, which leads to the oxidation of P 700, the so-called photoreaction I. He claimed furthermore a second photoreaction to refill the electron hole in the internal redox couples. We then would need a second sensitizer and a second donor which provides the required electrons. At EMERSON's time this seemed to be a very audacious hypothesis. DUYSENS could demonstrate that there really must be two light reactions, which he observed at the level of cytochrome *f*. The spectrum of this compound differs in the oxidized and the reduced state. DUYSENS realized that if he irradiated chloroplasts or whole cells with P 700 nm light he oxidized cytochrome *f*. If he took light between 660 and 680 nm he reduced the same compound [24]. This was the best evidence in favour of EMERSON's hypothesis.

Are the two light reactions working parallel or in series? Fig. 26 gives the present-day picture of the so-called *Z-scheme*. The first reaction, which is now called photoreaction I, is performed by a *photosystem I* (PS I); its donor is P 700; it transfers one electron to the primary acceptor, which might be a bound ferredoxin. From there it can be channelled to a chain of coenzymes until it finally ends on R—COOH, *i.e.* the primary fixation product of carbon dioxide.

The hole at P 700$^+$ will be primarily become filled up by cytochrome f and/or plastocyanin (see above). The postulated (second) *photosystem II* (PS II) should provide these redox couples with electrons. It must have its own sensitizers and its own donors and acceptors. It meanwhile turned out that chlorophyll a — bound to another macromolecule than the P 700 — acts as primary donor. By this special bond its absorption spectrum is shifted to the long wavelength side (however, much less than in the case of P 700). Not regarding the principal restriction in the application of the redox scale (see above) several physical chemists have tried to determine the redox potential of chlorophyll a *in vitro*. In organic solvents it seems to be between + 0.8 and 1.0 V [25]. It remains unknown, however, how this value must be taken *in vivo*. What is the first acceptor of PS II? If this would be the Q_1 itself, the first of the quinones, then there would be an energy difference between 0.8 and 1.0 V only. This, however, cannot be the case, for it is easy to demonstrate that illuminated chloroplasts can reduce artificial acceptors with a much more negative standard potential. During the last years Russian scientists presented evidence that the first acceptor in PS II might be *pheophytin, i.e.* a chlorophyll derivative in which the central magnesium atom is missing [26]. *In vitro* it has a standard potential of − 0.6 V. In its reduced state this primary acceptor would feed its surplus electron into the interconnecting chain of redox couples. Between the reduced acceptor of PS II and the oxidized donor of PS I (P 700$^+$) there is a span of nearly half a volt. From the energetic point of view it should be sufficient to produce ATP.

This simple scheme (Fig. 26) leaves many questions open. The critical problem how to refill the electron holes of the engaged redox couples is only shifted to the donor side of PS II. What happens with the oxidized donor of this system? In the meantime flash spectroscopical studies defined the time scale. They showed that the limiting step is behind the plastoquinone. It takes ∼ 20 ms to transfer an electron from plastoquinone to cytochrome f, whereas all other reactions run in the μs range. If we block this slowest step, as several inhibitors do, we get a delay in the whole photosynthetic process.

Fig. 27 summarizes the information, which immunologists contributed. They tried to study, which of the redox systems are located near the outer membrane side of the thylakoid and which are to be localized near the internal space. As far as our present-day information reaches, both donor systems, *i.e.* P 700 and P 680, are very near to the internal space, whereas the acceptors are near the outer medium (Fig. 27). So the electron transport has to bridge a distance of several nanometers. This creates a strong electrostatic field. If we draw this in terms of the redox scale we must anticipate that the standard potential of chlorophyll a will be near to

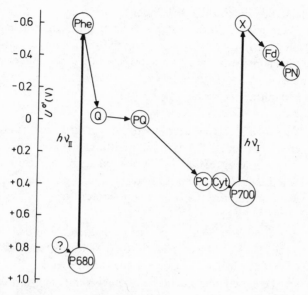

Fig. 26 — Light-induced electron transfer within the two photosystems of green plant cells (heavy arrows: light reactions, thin arrows: dark reactions). The exact standard potentials of the two electron acceptors (Phe: pheophytin, X: probably bound ferredoxin) and of the electron donor P 680 are still unknown. Q: unidentified quinone, PQ: plastoquinone, PC: plastocyanine, Fd: ferredoxin, Cyt: cytochrome f, PN: pyridinenucleotid (after Ref. 79, modified).

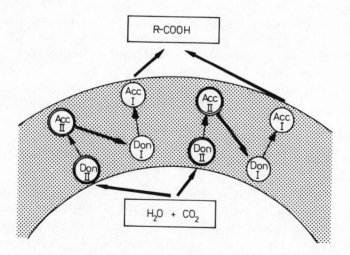

Fig. 27 — Localization of the donor and acceptor molecules of the two photosystems within the thylakoid membrane (after Ref. 80).

+ 0.8 V — at least not more than + 1 V. If the primary acceptor were slightly negative, like the first quinone, only less than 1 eV of the excitation energy (1.8 eV) would become converted into the redox potential difference. If, however, pheophytin (with its much more negative standard potential, see above) would be the first electron acceptor of PS II, then at least 1.4 eV would be recovered. The rest should be contained in the electrostatic field.

There are two possible back reactions to be regarded: one can (perhaps) lead to an ATP formation, but for PS II this is still questionable. The other back reaction can easily be discovered after switching off the exciting light: chloroplasts emit a weak but long-lasting *delayed fluorescence*. It has the same spectral characteristics as the spontaneous fluorescence, *i.e.* the energy content of the photons is 1.8 eV.

The energy difference between the excitation energy and the redox potential difference must be gained by the annihilation of the electrostatic field. This should contribute at least 0.4 eV, which is sufficient for the formation of an oxygen-phosphorous bond in ATP. There should be a mechanism which transforms the field energy into chemical energy of bonds. We do not yet understand how this functions, but there is convincing evidence that this transformation does not only happen in cyclic back reactions but also within the chain of redox couples spanning between the two photosystems (*non-cyclic photophosphorylation*).

In its reduced state acceptor PS I gives its surplus electron to fixed CO_2. We have to expect that the responsible enzyme system for this reaction — the ribulose-bisphosphat-carboxylase — lies near the outer surface. Indeed the electron microscope demonstrated *cobble stones* (see above) which apparently represent this enzyme. It remains a problem what happens *inside* the thylakoids. There is good evidence that illuminated thylakoids accumulate *protons*, respectively H_3O^+ cations [27]. Part of them are obviously *pumped* from the medium (Fig. 28), another fraction may originate in connection with the reactions on the donor side of PS II (see below). These are connected with the evolution of the photosynthetic oxygen.

11.   The source of photosynthetic oxygen

Since several decennia there are two contradicting hypotheses on the origin of photosynthetic oxygen. One theory — as favoured *e.g.* by WILLSTÄTTER [28] and later by WARBURG [29] — assumed that all the oxygen which is released by green cells is derived from CO_2, whereas most other laboratories presented evidence that it must come from water. Unfortunately we cannot work with radioactive oxygen isotopes — as many laboratories did so successfully with carbon isotopes. There are

Stroma

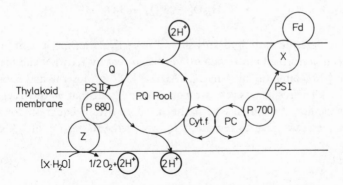

Internal space

Fig. 28 — Schematic diagram of the proton transport across the thylakoid membrane (*via* the PQ pool) and the additional proton production on the donor side of PS II. Z: Intermediate electron transfer system between (H_2O + CO_2) and P 680 (after Ref. 81, modified).

some radioactive carbon isotopes, but their lifetimes — except that of ^{15}O with ∼ 2 min — are only in the order of seconds or even microseconds. So they are too short to make them suitable tracers.

On the other hand there are three *stable isotopes*. The oxygen of the water and the atmosphere is a mixture of > 99 % ^{16}O with ^{18}O and smaller amounts of ^{17}O. In a mass spectrometer we can measure the weight of atoms and molecules, respectively of ions and radicals. In principle it should therefore be possible to work with ^{18}O-labelled photosynthesis substrates (water or CO_2). With these tests we could hope to decide which substrate is the real source of the photosynthetic oxygen.

If we supply cells with labelled water, they should incorporate its oxygen into the reaction products of the photosynthetic process. If WARBURG were correct, then all ^{18}O should become incorporated into the synthesized carbohydrates, whereas the released oxygen is derived from the unlabelled carbon dioxide:

$$6\,CO_2 + 6\,H_2O \xrightarrow{\;n\,h\nu\;} C_6H_{12}O_6 + 6\,O_2 \tag{16}$$

Thus if we determine the molecular weight of the photosynthetic oxygen it should be 32. If the other laboratories were correct, then we should observe that 50 % of the evolved oxygen is ^{18}O, the other half ^{16}O:

$$6\,CO_2 + 6\,H_2O \xrightarrow{\;n\,h\nu\;} C_6H_{12}O_6 + 3\,O_2 + 3\,O_2 \qquad (17)$$

In this case the carbohydrates should have their normal weight. This experiment has been performed more than 40 years ago first by RUBEN and his coworkers in Berkeley [30], later on by VINOGRADOV [31] in Moscow and some Japanese scientists [32]. The result was very strange: it turned out that *all* the oxygen released is ^{18}O. We cannot explain this with the usually written equation for the photosynthetic process; we have to rewrite it following the suggestion of VAN NIEL (see below):

$$6\,CO_2 + 12\,H_2O \xrightarrow{\;n\,h\nu\;} C_6H_{12}O_6 + 6\,O_2 + 6\,H_2O \qquad (18)$$

Then we end up with normal carbohydrates and 6 *new* water molecules, which are the product of secondary reactions — not to be mixed with the supplied labelled water! Under this condition all the oxygen can be ^{18}O.

Since that time most biochemists and plant physiologists are convinced that oxygen is formed from water. That would mean that we have to decompose H_2O. Everybody must be aware that the splitting of an O–H bond requires much more energy (~ 4.8 eV) than contained in a photon of red light (1.8 eV). So we face first a difficult energetic problem. We have to ask whether *free* water is the actual reaction partner. Could there perhaps be a type of *bound* water with quite different electrochemical properties?

In this context another experiment has to be mentioned: in photosynthetic bacteria it was shown that their cells use H_2S as electron donor [33]. VAN NIEL, a Dutch microbiologist, observed already in the thirties that if we supply suspensions of photosynthetic bacteria with hydrogen sulfide, we obtain carbohydrates and at the same time deposited elementary sulfur. So VAN NIEL wrote a very general formula:

$$6\,CO_2 + 12\,H_2X \xrightarrow{\;n\,h\nu\;} C_6H_{12}O_6 + 6\,H_2O + 12\,X \qquad (19)$$

He regarded that in the case of bacteria X is identical with sulfur, in the case of green plant cells with oxygen. It is, however, not justified to compare these two systems: In a mixture of CO_2 and H_2S there is no chemical reaction; with CO_2 and H_2O we have on the other hand the formation of bicarbonate and carbonate anions:

$$CO_2 + 3\,H_2O \rightleftarrows HCO_3^- + H_3O^+ + H_2O \rightleftarrows CO_3^{2-} + 2\,H_3O^+ \qquad (20)$$

The system of the higher plant is much more complicated: we know that the equilibrium between CO_2 and H_2O is very efficiently regulated by a special enzyme, the carbonic anhydrase which has the highest turnover rate of all known enzymes [34]. This makes it very difficult to come to a discrimination between the two substrates. In all physiological systems we have a ratio of 1 CO_2 to $\sim$ 100 000 water molecules. Whenever we label 100 000 H_2O and leave 1 CO_2 unlabeled, then — after equilibration — everything will be labeled. If we, on the other hand, label the CO_2 and suspend the cells in normal water, the dilution effect is so extreme, that we will not find any labeled atom in the reaction product. This is a serious argument against RUBEN's and VINOGRADOV's experiments [35].

These are two arguments in favour of a water decomposition hypothesis. There is a third one: the once believed assumption that we need no CO_2 for oxygen evolution with isolated chloroplasts. Since the beginning of this century plant biochemists and biologists had repeatedly tried to isolate plastids and to observe the complete photosynthetic process in organell preparations. They were never successful. These experiments were resumed by ROBERT HILL in Cambridge (1935). He anticipated that the light-induced processes start with a charge exchange, and he considered during the isolation procedure either donor or acceptor might go lost. He first tried to supply his preparations with an artificial electron acceptor; for this purpose he chose ferric ions. So he started with the (unknown) natural donor of the chloroplasts — probably the chlorophyll P 680 (see above) — and an artificial donor.

By irradiation he expected to obtain an oxidized donor and ferrous ions:

$$\text{Don} + \text{Fe}^{3+} \xrightarrow{\ h\nu\ } \text{Don}^+ + \text{Fe}^{2+} \tag{21}$$

Actually he observed a constant stream of molecular oxygen [36]. He knew nothing about the nature of the internal donor; but if a secondary reaction led to molecular oxygen, he dared to say that the oxidized donor decomposes water. So its electron affinity must be high enough to remove one electron from H_2O. This means to postulate oxidized water:

$$\text{Don}^+ + H_2O \longrightarrow \text{Don} + H_2O^+ \tag{22}$$

At the time of HILL's first experiments nothing was known on this radical. Today we know it from radiochemical experiments. It has an extremely short lifetime, in aqueous systems $\sim 10^{-13}$ s. The reaction which happens with unchanged water molecules is a proton exchange:

$$H_2O^+ + H_2O \longrightarrow H_3O^+ + OH^\bullet \tag{23}$$

By this step we would obtain H_3O^+, so-called hydrogen ions, and $OH^\bullet$-radicals. HILL assumed that there must be a pathway from $OH^\bullet$ radicals to molecular oxygen

$$2\ OH^\bullet \rightarrow \rightarrow \rightarrow\ O_2\ +\ ? \tag{24}$$

The same secondary reaction sequence should be part of water electrolysis. Here we transfer OH^- anions to the anode where they deliver one electron to the metal surface. We observe that the resulting $OH^\bullet$ radicals are transformed into O_2 molecules. If we look into the physicochemical literature we find to our surprise that nobody can give us proven details on the intermediate states [37].

HILL's experiments were later repeated and extended by OTTO WARBURG with some other electron acceptors, like p-benzoquinone. He observed that if he removed all the CO_2 from the chloroplast suspensions there was no oxygen evolution [38]. Many laboratories meanwhile reproduced and could confirm WARBURG's tests. Without carbon dioxide the HILL reaction does not proceed. On the other hand we need only very small CO_2 concentrations. The amount of oxygen which is produced is > 100 times higher than the added carbon dioxide. So we may perhaps call this effect a *catalytic* influence of CO_2 [39]. At any case we cannot say that HILL reactions work *without* carbon dioxide.

Biochemical and biological textbooks usually compare the situation in PS II with electrolytic water decomposition. For this they claim a necessary potential of 1.23 V, corresponding to the potential difference between the hydrogen and the oxygen electrode. This is wrong! This value is valid for a *reversible* reaction; an electrolytic decomposition is an *irreversible* process which needs much more energy. In fact, nobody has ever seen an electrolytic H_2O cleavage with less than 1.6 V. If the primary donor P 680/P680$^+$ has a standard potential of $+ 0.8$ V, it is very difficult to grasp how electrons from water should refill the hole. In physiological systems we often use to compare the standard potentials with that of the oxygen electrode. This is the potential of the couple $O^{2-}/O^\bullet$, *i.e.* a *two*-electron step. In HILL reactions we assume that there is only a *one*-electron transfer [according to equation (22)]. The sequential removal of the two electrons from O^{2-} does not cost identical energies [40]; it takes much more energy to withdraw the second one. To oxidize water to H_2O^+ we need at least 2.3 eV, but in (red light) photons only 1.8 eV are delivered to the antennae. So we have to postulate an additional mechanism to explain this discrepancy.

We owe valueable informations to experiments of JAMES FRANCK [41] and later on of JOLIOT and KOK [42]. FRANCK tried to get oxygen by single light flashes, but he could not observe any oxygen release by only one flash. Fig. 29

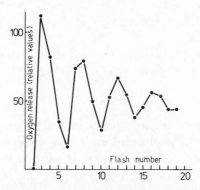

Fig. 29 — Oscillations of oxygen release in a series of (μs) light flashes as function of flash number. Details see text; (after Ref. 82, modified).

demonstrates a very interesting observation of JOLIOT and KOK. It gives the oxygen yield of very short (μs) light flashes as function of the flash number. We see that the first flash releases no oxygen at all. There is sometimes some oxygen after the second flash, whereas the third flash leads to a high O_2 yield. Extending the flash sequence we obtain a pattern which looks like a damped *oscillation* with a period of four. The easiest interpretation would be to claim four additive quantum absorptions, by which four photons, each with 1.8 eV, are *collected* somewhere. Then their combined energies would be sufficient to split water. A hypothetical scheme might be written as

$$S_0 \xrightarrow{h\nu} S_1 \xrightarrow{h\nu} S_2 \xrightarrow{h\nu} S_3 \xrightarrow{h\nu} S_4 \tag{25}$$

The (chemically unidentified) intermediates are called *S states*. This is very unfortunate, because this can be mixed up with the S (singlet) states as reserved for the description of energy terms. The lowest state would be S_0, but we have to postulate that after extended dark times the somewhat *energized* state S_1 is the dominating species (otherwise it cannot be understood why already the third flash gives oxygen). Nobody knows what these S states could be; some are inclined to identify them with different oxidation states of a still unknown redox system. They would write a series like:

$$S \xrightarrow{h\nu} S^+ \xrightarrow{h\nu} S^{2+} \xrightarrow{h\nu} S^{3+} \xrightarrow{h\nu} S^{4+} \tag{26}$$

The last intermediate (S_4) reacts with H_2O to give O_2 and four protons:

$$S^{4+} + 2 H_2O \longrightarrow S_0 + 4 H^+ + O_2 \qquad (27)$$

Apparently S_0 and S_1 are more or less stable states, whereas S_2 and S_3 have a lifetime in the order of seconds. Only the highest oxidized state (S_4) is very unstable.

How can quantum energies become stored? In plant physiology it is known since a long time only cells provided with *manganese* can release oxygen. If this element is removed the O_2 evolution stops. It has been convincingly demonstrated that there are two types of *bound* manganese [43]. One, which is very loosely fixed, can be washed away from thylakoids by Tris, the other one — about 1/3 the metal content — is very tightly bound. Each reaction center contains apparently 6 manganese atoms. From inorganic chemistry we know that manganese is an element with many oxidation states. It is not unconceivable to assume that the S states represent different oxidation states of a complicated manganese comples. Several laboratories have meanwhile tried to isolate manganese-containing proteins. There are some *identifications* in the literature, which have, however, not been reproducible until now.

S states do not explain our main problem: if all these states have to feed their electrons into the holes of oxidized P 680, their standard potentials cannot reach values which are positive enough to oxidize H_2O. So there remains the problem: do we really decompose *free* water or do we have a type of *bound* water with completely different thermodynamic properties? The data for *bound* CO_2, *i.e.* of the R–COOH (see above), are quite different from those of the free molecule. So it would be conceivable that *bound* water can be more easily oxidized than pure H_2O.

There are, indeed, some indications that there is something like *bound* water: If we wash cells with $H_2 {}^{18}O$ we would expect a fast exchange between the *normal* water inside the cell and the *heavy* water in the medium. But KUTYURIN could demonstrate that part of the cell water cannot be exchanged [44]. So a certain fraction seems to be tightly bound; we do not know, in which form.

Another explanation would be to assume that it is not the water which is decomposed, but something else. WARBURG called the hypothetical precursor of the photosynthetic oxygen a *photolyte* [45] — and he assumed that it would be identical with a *functional carbonic acid* which he gained from cells by treatment with fluoride. Not regarding this very special claim, the *photolyte* could well be a compound, in which both water and CO_2 are bound. We can rewrite the photosynthesis equation once more, as proposed by VAN NIEL:

$$6 \, CO_2 + 12 \, H_2O \xrightarrow{\; n \, h\nu \;} C_6H_{12}O_6 + 6 \, O_2 + 6 \, H_2O \tag{28}$$

We know that we obtain an equilibrium between the two photosynthesis substrates:

$$6 \, CO_2 + 12 \, H_2O \rightleftharpoons 6 \, HCO_3^- + 6 \, H_3O^+ \tag{29}$$

We have to ask which side of this equation is predominating in biological systems. If we consider the hydrogen ion concentration inside the chloroplasts, we face a complicated situation: apparently the plastid stroma — the place where the carbohydrates are synthesized — has a pH value of at least 8, whereas inside the thylakoids, where the oxygen is produced, the value might be near 5 [46]. Therefore on both sides of the thylakoid membrane different CO_2/HCO_3^- ratios should occur. Obviously the CO_2-fixing enzyme, the ribulosebisphosphate-carboxylase (see above) needs the CO_2 in spite of the better availability of the HCO_3^- anion. Nobody can say how the situation on the other side of the membrane must be; it cannot be excluded that the bicarbonate anion plays a role in the oxygen-releasing partial reaction.

Do we really decompose oxygen-hydrogen bonds or is there any evidence that another type of bond is broken? This question can be studied with two different methods: one could try to develop inorganic model systems which simulate HILL reactions and show that under reproducible conditions all the steps between H_2O and CO_2 on one side and molecular oxygen on the other side can be studied on an electrode or salt surface. The other way would be to perform new isotope experiments. Discussing RUBEN's experiments we realized how difficult experiments of this kind must be, due to the very fast isotope exchange between water and CO_2. If we introduce an unusual oxygen isotope into either H_2O or bicarbonate, this is within a few seconds distributed all over the system. So we have to look for some *tricks*.

The CO_2 share of the atmosphere is just 0.03 %. This caused a long discussion in biochemistry, whether this low concentration could be sufficient to drive the CALVIN cycle, especially whether it would be enough for the first enzyme, which has to fix the carbon dioxide. Looking into the literature we find that WILLSTÄTTER already (1920) observed that the CO_2 concentration in some plants is much higher than to be assumed [47]. Starting with an *external* concentration of 0.03 % we would expect *internal* concentrations in the μmolar range. WILLSTÄTTER observed that in some leaves concentrations up to 100 mM occur. So there must be a mechanism to *concentrate* the carbon dioxide. When WARBURG observed that we need CO_2 for HILL reactions, he tried to determine its internal concentrations. He released the *bound* carbon dioxide by the addition of fluoride anions. Interpreting his measurements he claimed that for each chlorophyll molecule there is one

CO_2. This stoichiometric ratio caused biochemists to look for a chlorophyll-CO_2 complex, but nobody could ever find it. With isolated chloroplasts we cannot confirm this strict stoichiometric ratio; different from whole algal cells, chloroplasts contain one CO_2 for ~ 5 chlorophyll molecules. We do not yet know where to localize it. If we destroy the proteins of the thylakoids — by treating the suspensions with various proteases — the concentration of the bound CO_2 remains unchanged. If we, on the other hand, dissolve part of the lipids by lipases, the concentration of the internally bound carbon dioxide decreases significantly.

For the formation of sugars we need an acceptor which itself is already a sugar (ribulosebisphosphate). After extended dark times we would expect that the cells start with a very low acceptor concentration. At the beginning of the light period the photosynthesis rate should therefore be low. It is not before the CALVIN cycle is running, that the acceptor concentration can increase.

Fig. 30 demonstrates the actual determination of the CO_2 uptake. We observe that within a very short time — less than 30 s — a high amount of CO_2 is taken up. If we switch off the light, CO_2 is released; under certain experimental conditions this might be the same amount as that which had been formerly fixed. So part of the CO_2 binding must be light-dependent. Japanese authors could demonstrate that WARBURG's *functional carbonic acid* is not channeled into the CALVIN cycle. If the two phenomena are related, there must be an unknown form of bound CO_2. This is apparently the carbon dioxide which WARBURG removed by the treatment with F^- anions. In the meantime we know that there are other anions — *e.g.* acetate and formate — which release this bound CO_2 [48]. The CO_2-depleted systems are unable to perform HILL reactions. We have to add catalytic amount of bicarbonate to reactivate them.

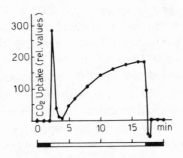

Fig. 30 — CO_2 uptake after an extended dark period (so-called *induction*). Data taken from observations on moss leaves (after Ref. 83, modified).

Is it possible to perform the whole sequence of a HILL reaction under completely artificial conditions — *i.e.* without any biological material? When we started experiments on models we anticipated a water oxidation. We thought that we only have to provide a suitable electron donor which in its oxidized state can remove electrons from H_2O molecules, *i.e.* we have to have a species with a very high electron affinity. This is an electrochemical problem. We can look into appropriate

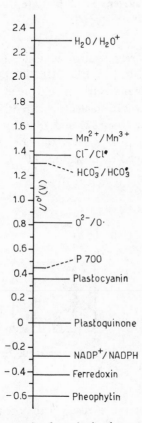

Fig. 31 — Scale of standard redox potentials.

data collections to decide what to use. Fig. 31 gives an energy scale with the oxygen electrode (at pH 7) at a potential of + 0.82 V. To oxidize water we should have a redox system with a more positive standard potential. At first sight it looked as if we could take chloride anions. The mixture $Cl^-/Cl^\bullet$ has a potential of + 1.36 V. So we searched for a photochemical process to produce chlorine radicals. A reaction of this kind is the *photographic process* [49]. The basic step is the transfer of one electron from a halogen anion — *e.g.* Cl^- — to silver cations:

$$Ag^+ + Cl^- \xrightarrow{h\nu} Ag^0 + Cl^\bullet \tag{30}$$

We should end up with metallic silver and chlorine radicals. In the photography we are only interested in the metal. What we do not see is the chlorine, because it is adsorbed by the gelatine of the film. If we irradiate AgCl in a closed vessel and introduce the volatile reaction products into a mass spectrometer we can easily detect the released halogen. So it is quite obvious that we obtain chlorine gas.

Our first tests were purely electrochemical experiments. They made use of a well-known effect first published by BECQUEREL [50]. This author described that whenever we cover two metal electrodes with a film of a suitable dye and then irradiate one of them, we observe an electric current. We replaced the formerly used amperemeter by a voltmeter (Fig. 32) and tried to see whether there is any such BECQUEREL *effect* with sensitized silver chloride. We covered silver rods electrolytically with AgCl and sprayed them with a chlorophyll solution. By irradiation of

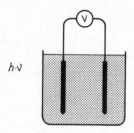

Fig. 32 – Simple set-up for the measurement of the BECQUEREL effect on semiconductor electrodes (Details see text).

one of the two electrodes we obtained a photo-potential with the opposite sign as to be expected from BECQUEREL's data. In his experiments all metal electrodes attained a positive charge, in our experiments [51] the irradiated semiconductor AgCl became negatively charged — even in contact with distilled water (*i.e.* a saturated solution of the very sparely soluble silver chloride). We thought that we would end up — as in the HILL reaction — with H_2O^+. This is a very short-living species, which within 10^{-13} s exchanges a proton with a water molecule (see above), so that we should expect $OH^\bullet$ radicals:

$$H_2O^+ + H_2O \longrightarrow H_3O^+ + OH^\bullet \tag{31}$$

The electrodes of our experimental set-up had a surface of only a few square millimeters. It can easily be calculated that the amount of possibly resulting oxygen molecules would be too small to become detectable by ordinary oxygen electrodes.

We can replace the electrode surface by a fine crystal powder. By mixing a cold silver nitrate solution with HCl and lyophilizing the precipitate we gain preparations with specific surfaces of more than $1 \, m^2$ per gram. If we distribute and irradiate this material in water we observe, as in HILL reactions, a constant stream of oxygen [52]. We may assume that the formed chlorine radicals regain their missing electrons from water.

$$Cl^· + H_2O \longrightarrow Cl^- + H_2O^+ \qquad\qquad (32)$$

This would give the same radical as in the case of isolated chloroplasts — and probably the same pathway from $OH^·$ radicals to O_2. Nobody has, however, seen hydroxyl radicals in photosynthetic systems. When we observed this process with freshly precipitated silver chloride, we realized from day to day a fluctuating quantum efficiency. After a while we recognized that preparations after some *waiting time* often gave a much better yield. It seemed possible that first CO_2 of the surrounding air has to become absorbed by the suspension, *i.e.* that a certain bicarbonate concentration has to be reached before the system can work. It is very easy to test this assumption by the injection of bicarbonate. In fact HCO_3^- addition increases the quantum yield considerably [53].

After this unexpected observation it was tempting to perform experiments on the source of the released oxygen. Is this really the water or could it perhaps be the bicarbonate? Fortunately enough we had the opportunity to make these experiments in the Nuclear Research Center at Cadarache (France), where a sensitive mass spectrometer was available. For these tests ^{18}O-labelled bicarbonate was taken. During irradiation the evolved oxygen turned out to be significantly enriched in ^{18}O [54]. In the case of AgCl we may therefore say that the O_2 comes from the bicarbonate, not from the water. This, of course, means nothing for the natural system. Facing the problem of the very fast oxygen isotope equilibration between H_2O and CO_2 it seems at first sight completely impossible to decide which is the real precursor of the photosynthetic oxygen. Realizing the limited value of former experiments by RUBEN and many others there remains for studies with biological samples only the possibility to look for material free of carbonic anhydrase. Even then only short illumination periods should be chosen because the isotope equilibration runs even without any catalyst.

With a very precise instrument (which detects small deviations in the $^{18}O/^{16}O$ ratio) it is not absolutely necessary to label the substrates at all. Natural water is a mixture of H_2O and D_2O. These two species differ in the only chemical bond

(H—O, respectively D—O). The zero point energies of these two bonds are different (Fig. 33). So also the activation energies for their breakage must differ. There is a smaller energy demand for the splitting of the oxygen-hydrogen than for that of the oxygen-deuterium bond. What is the consequence? If we decompose water electrolytically the probability to break the O—H bond is somewhat higher than that for the O—D bond. This may cause rate differences by a factor of > 50 [55]. We call this a *kinetic isotope effect*. So we understand why by electrolytic processes D_2O compared to H_2O can be enriched.

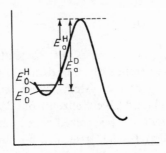

Fig. 33 — Zero point (E_o) and activation energies (E_a) for the breakage of O—H and O—D bonds. The scheme demonstrates the lower energy requirement for the splitting of O—H bonds (after Ref. 84, modified).

Water contain also $\sim 0.2\ \%\ H_2{}^{18}O$. Due to the smaller mass ratio ($^{18}O/^{16}O$ compared to $^2H/^1H$) the isotope discrimination between the oxygen isotopes should be much smaller than between the hydrogen isotopes. But nevertheless the ^{16}O—H bond should be somewhat easier broken than the bond between ^{18}O and hydrogen. Between water, CO_2, bicarbonate and hydrogen ions a temperature-dependent equilibrium is established:

$$CO_2 + 2\ H_2O \rightleftharpoons HCO_3^- + H_2O^+ \tag{33}$$

Due to slightly different activation energies the rates for forward and backward reactions are different for the two isotopes ^{16}O and ^{18}O. This leads to an *equilibrium isotope effect* resulting in an enrichment of the heavier isotope in the heavier species. So both CO_2 and bicarbonate anions have a definitely higher ^{18}O content than the water (and its ion). Table 1 gives the data for 25 °C.

In fresh water the ^{18}O content is between 0.198 and 0.199 %. Since the *normal* water has a somewhat lower boiling point than the ^{18}O-labelled water, the evaporation enriches the heavier species significantly. In ocean water we therefore observe a higher ^{18}O content (0.1995 %).

Table 1. — Relative ^{18}O content in the compounds of the H_2O–CO_2 system [56]

^{18}O Content		^{18}O Content	
H_2O	1.000	HCO_3^-	1.026
H_3O^+	1.000	CO_2	1.041

The measurements demonstrated that the ^{18}O content in the bicarbonate anion is 2.6 % higher than in the water. So by this *equilibrium isotope effect* we gain substrates with distinguishable isotope ratios, which are no more modified by any exchange reaction. This opens a promising way for experiments with the natural isotope abundances.

Since it is easier to break a bond between ^{16}O and hydrogen than between ^{18}O and hydrogen, we should expect that the oxygen which is released during water splitting is enriched in the lighter (normal) isotope. As to be seen from Table 2 the normal ^{18}O content of fresh water is 0.1981 %. In electrolytic decompositions the ^{18}O content of the released O_2 is decreased; the exact $^{18}O/^{16}O$ ratio depends on several parameters, *e.g.* the kind of electrode material and the current density. If photosynthesis would also include a splitting of an oxygen-hydrogen bond, we should expect a similar effect. We observe just the reverse: the enrichment of the heavier isotope. This is a very unexpected result.

Table 2. — ^{18}O content of oxygen samples evolved from fresh water (after Ref. 57)

	^{18}O Content (%)	^{18}O Content (relative)
Fresh water	0.1981	1.000
Electrolytic oxygen	0.1961	0.990
Photosynthetic oxygen:		
Elodea leaves	0.2003	1.011
Land plants	0.1992	1.006
Fresh water algae	0.1991	1.005

It cannot be excluded that this shift in the $^{18}O/^{16}O$ ratio is caused by cell respiration. We may suppose that the respiring cell prefers the lighter isotope and leaves the ^{18}O behind. The data on isotope discrimination rates by heterotrophic cells from bacteria till to human tissues show, however, too small differences. Nevertheless several authors tried to explain the observed ^{18}O enrichment in the photosynthetic oxygen by an isotope effect of mitochondrial respiration [58]. At any case we may expect more convincing results if it would be possible to exclude this pretended interference.

If a suspension of chlorophyll-containing cells or isolated chloroplasts (thylakoids) in an aqueous solution is irradiated inside a closed vessel, the photosynthetic O_2 can be collected in a pre-evacuated sampling glass. For experiments lasting only a few minutes we need a very sensitive instrument, because — quite different from the test performed by RUBEN or by VINOGRADOV — we cannot expect to collect several cubic centimeters of gas, but just a few cubic millimeters. To avoid the enzymatic enhancement of the H_2O–CO_2 exchange the first experiments were performed with cells [59], which had no carbonic anhydrase.

Much more convincing are experiments with preparations in which there is no respiration at all. Respiration and photosynthesis are bound to different cell organells: all processes which consume oxygen are linked to mitochondria (or to peroxysomes) whereas the photosynthesis process is restricted to the chlorophyll-containing plastids (chloroplasts). With very good (mitochondria-free) plastid preparations — gained e.g. by centrifugal fractionation of cell homogenates — we have no light-dependent oxygen consumption any more. The only pitfall could be a photooxidation, the influence of which has to be excluded by suitable control experiments. So we had to repeat the former experiments on cells with chloroplasts or thylakoids. We obtained the same result [60]. This indicates that there is — independent from respiration — an oxygen isotope discrimination which in its sense differs from the isotope effect in all inorganic systems.

It could be argued that it is not *free* water which is decomposed, but a kind of *bound* water with different properties. To test this possibility we had to make control experiments. How does photosynthesis work in D_2O? In fact such tests have been made in 1935 already. Quite independent from each other two laboratories came to the same result: if we replace normal water by D_2O, we measure a — by a factor of 2-3 — lower photosynthesis rate [61, 62]. In the context of oxygen evolution this does not mean anything, because we do not know which of the many partial reactions discriminate between the two hydrogen isotopes. We have to choose reaction conditions in which the oxygen evolution becomes rate-limiting. This is rather easy to achieve: under the conditions of very dim light the photon supply becomes the decisive factor. So we repeated the experiments first at very low light intensities, then with very short (μs) light flashes, alternating with (up to 100 ms) long dark periods. In this case there is practically no discrimination between hydrogen and deuterium. We face a very curious situation: the living plant apparently discriminates between oxygen but not between hydrogen isotopes. The easiest conclusion from this observation would be to postulate that the bond which is decomposed contains oxygen but not hydrogen. So it cannot be an O—H bond. The sensitive bond must either be a C—O or an O—O (*e.g.* a peroxide) bond. We cannot decide this alternative just now, because there are no experimental data with ^{13}C. At any case it means: we do not split water.

In the old photosynthesis literature we find both in WILLSTÄTTER's and in FRANCK's publications good arguments that there are intermediates of peroxidic nature. If we irradiate suspensions of chlorophyll-containing cells or of chloroplasts there is, indeed, a fast formation of peroxides. It could well be that their O—O bond becomes decomposed. The problem mentioned already is that one quantum of red light (containing 1.8 eV) has not enough energy to split an O—H bond (which requires $\sim$ 5 eV). This difficulty would disappear if we have to assume that we decompose an O—O bond (which would ask for $\sim$ 1.6 eV).

The results of experiments with unlabelled substrates are insufficient to justify such far-reaching conclusions. We were well aware that we had to perform new experiments with labelled CO_2 or water. To avoid the catalyzed isotope equilibration (see above) we obviously have to remove the decisive enzyme. For animal cells we know several inhibitors which really stop the action of the carbonic anhydrase. But we cannot apply these compounds with the same success in plant cells. With tolerable concentrations they block the enzyme only to $\sim$ 30-40 %. Fortunately enough there are indeed organisms (blue-green algae) without carbonic anhydrase. Furthermore we can choose an isolation procedure for chloroplasts so that the finally obtained thylakoid preparation is free of this enzyme. This opens a way for reliable experiments with labelled photosynthesis substrates — provided the available mass spectrometer has the necessary sensitivity to measure cubic millimeter volumes with the sufficient accuracy.

With organisms free of carbonic anhydrase we can apply ^{18}O-labelled substrates. Unfortunately plant cells contain a high amount of *bound* CO_2 (see above), which severely dilutes the heavy isotope of the externally supplied carbon dioxide. Even thylakoids possess this fraction, but in this case at least a considerable part of it can be removed. For this purpose the isolated organells are washed with formate solutions (see above). After this pretreatment they have no photochemical activity any more. This can, however, be restored by the injection of either $NaHC^{16}O_3$ or $NaHC^{18}O_3$. After irradiation we can collect the released oxygen and determine its $^{18}O/^{16}O$ ratio. The samples provided with the unlabelled salt serve as controls; the relevant data may be expected from the suspensions supplied with $NaHC^{18}O_3$.

If water should be the real oxygen precursor, we obviously start with an unlabelled oxygen source and should accordingly obtain an unlabelled product. If, on the other hand, the O_2 has to be derived from the (primarily highly labelled) bicarbonate the released O_2 must also be labelled. In fact we observe — both for intact cells and for isolated thylakoids — that the *early* oxygen (collected up to 60 s) is significantly enriched in the heavier isotope [63].

Compared to intact cells thylakoids we have an additional advantage: they have no respiration (see above). There may be some photo-oxidation but this interference can easily be considered by suitable control experiments.

With any mass spectrometer very small gas volumes may cause severe artifacts. These include not only fractionations by different solubilities or adsorption of the two isotopes, but also a discrimination at any porous membrane. For this reason it is dangerous to evaluate the directly measured single curves. It is more reliable to consider *difference curves*. In their traces the conditions for parallel experiments are identical — except the $^{18}O/^{16}O$ ratio of the applied bicarbonate. Fig. 34 gives exactly the result which was formerly obtained for the (carbonic anhydrase-free) blue-green algae.

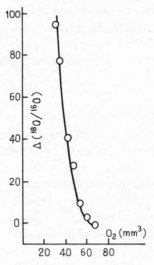

Fig. 34 — $^{18}O/^{16}O$ ratio of released photosynthetic oxygen as function of the evolved gas volume. The difference curve (details see text) was calculated for suspensions of isolated thylakoids.

These data show that the bicarbonate oxygen is incorporated into the photosynthetic oxygen. On the other hand we had seen from HILL reaction experiments, that we cannot expect to split just CO_2 as originally postulated by OTTO WARBURG. The easiest interpretation would be that there is a *cyclic process*. The first reaction of CO_2 after its entrance into a cell is its interaction with water [see equation (33)], which gives HCO_3^- and hydrogen ions. As working hypothesis we may assume that the oxidized donor of photosystem II does not react (as postulated in HILL reactions) with water, but with the HCO_3^- anion:

$$Don^+ + HCO_3^- \longrightarrow Don + HCO_3^* \tag{34}$$

We then would end up with bicarbonate radicals; at the same time we obtain the re-reduction of chlorophyll which was primarily oxidized. From the energetic point of view this would be much easier, because it costs much less energy to remove an electron from a negatively charged anion than from a neutral molecule.

What would happen to the HCO_3^* radicals? We first thought that they would be decomposed into OH^* radicals and CO_2:

$$HCO_3^* \longrightarrow CO_2 + OH^* \tag{35}$$

In the early interpretation of the HILL reactions we always anticipated to get OH^* radicals. We normally assume that these combine to form molecular oxygen (see above).

It seems unlikely that there are OH^* radicals in photosynthetic cells. To avoid further speculations on the fate of the HCO_3^*, this radical can be produced by a simple procedure. In fact it is rather easy to produce it electrolytically. If we electrolyze a bicarbonate solution, the anions deliver one electron to the anode and so become oxidized to the corresponding radical. With a very big cathode and a very small anode the current density near the anode becomes very high. This should keep the local concentration of the radicals rather high, too. Apparently there is no decomposition, but the formation of a dimer,

$$2\,HCO_3^* \longrightarrow HOOC-O-O-COOH \tag{36}$$

the *peroxidicarbonic acid*. Strangely enough this compound has a nearly insoluble potassium salt [64]. So by electrolysis of $KHCO_3$ solutions we easily obtain gram quantities of its salt.

Chemists have meanwhile studied various peroxi-acids. They know that these compounds are rather labile. Peroxidicarbonic acid decomposes into water, CO_2 and oxygen:

$$HOOC-O-O-COOH \longrightarrow 2\,CO_2 + H_2O + O^\bullet \qquad\qquad (37)$$

In the presence of unsaturated molecules — like *e.g.* carotenoids — they transfer oxygen to C=C double bonds. *In vitro* (some) carotenoids are converted into their epoxides. Plant physiologists observed that in normal photosynthesis cells always produce carotenoid-epoxides. Nobody knows the sense of this side reaction. At the present state of knowledge it would be premature to claim that it is part of the normal pathway to oxygen.

The resulting CO_2, on the other hand, would react with H_2O and by this means restore the secondary donor bicarbonate. There would really be a *catalytic* effect. It would explain why we do not *consume* but always *recycle* the CO_2. There remains certainly a long way to test the validity of this concept. The CO_2 effect must not be ascribed to the free HCO_3^- ion. It is possible that there is a complex between the anion and an as yet unidentified molecule. So we prefer to write $[X^\bullet HCO_3]^-$. This would probably also produce a bound intermediate.

12. Artificial systems

It was already mentioned that there are several redox couples which can replace the natural electron acceptor(s). Part of them were used *e.g.* by OTTO WARBURG and ROBERT HILL as HILL reagents. This demonstrates that we can drain off electrons from their normal channel. It could be interesting to have artificial systems which produce *molecular hydrogen*. With a strongly reducing agent it should be possible to reduce H_3O^+ ions (Fig. 35). There are indeed many microor-

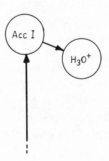

Fig. 35 — Scheme for the (dark) electron transfer from the reduced primary acceptor of PS I to a hydrogen ion. This step starts a reaction sequence which leads to the production of molecular hydrogen (details see text).

ganisms which can transfer electrons to hydrogen ions. They possess special enzymes, the *hydrogenases*. There are no hydrogenases in chloroplasts, but we can set up a very simple system by mixing isolated chloroplasts or thylakoids with hydrogenases, which were previously gained from mass cultures of bacteria. If we irradiate them, we observe that the electrons do not (all) flow into the CALVIN cycle but that they are taken for the production of molecular hydrogen. Hydrogenases are anaerobically working enzymes. In the presence of O_2 their activity is depressed. So by the concurrent oxygen evolution the reaction $H_3O^+ \rightarrow H_2$ stops after a short time.

There is, however, an organism (*Desulfovibrio desulfuricans*) which has a hydrogenase which seems to be insensitive against O_2. With enzymes isolated from this microorganism it should be possible to produce hydrogen even when the complete system is running. Otherwise we would have to interrupt the electron pathway by suitable inhibitors which prevent the evolution of oxygen.

This system might not be very interesting for practical use: The energy density of solar radiation is pretty low if we compare it with other energy sources. So we would have to collect the hydrogen over very vast areas. This is difficult because H_2 permeates most transparent materials. Another objection would be that biological systems are very labile; they survive only up to a few hours. Some groups therefore tried to replace isolated chloroplasts and even enzyme solutions by platinum suspensions [65]. By this means they obtained artificial systems which are indeed able to develop H_2.

There is another interesting artificial system, which works on the other end of the electron transport chain. It turned out that we can *feed* electrons into photosystem II from inorganic sources like Mn^{2+} cations [66]. Fig. 36 sketches a very

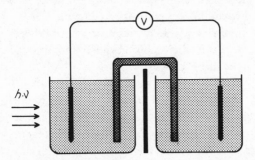

Fig. 36 — Simple rechargeable *battery*. Both beakers (connected by a KCl bridge) filled with a chloroplast suspension in Mn^{2+}-containing medium. In the illuminated (left-hand) compartment the manganese is oxidized, in the dark (right-hand) control vessel the oxidation state remains unchanged. A voltmeter gives the resulting potential difference.

simple electro-chemical system: two glass beakers filled with a suspension of unicellular organisms or chloroplasts. Both contain Mn^{2+} in the form of manganese chloride, both are equipped with a platinum electrode. If we irradiate one of the two compartments we oxidize Mn^{2+} to Mn^{3+} (Fig. 37), whereas in the dark compartment the manganese remains in the less oxidized form. So we obtain a *battery* which charges itself by photons. This is another system with which we can simulate parts of the photosynthesis process.

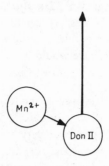

Fig. 37 — Re-reduction of the oxidized donor of PS II by Mn^{2+} ions. This well-established redox reaction is hardly understandable under the assumption of a standard potential of + 0.8 V for the primary electron donor.

Even more challenging might be the effort to reduce CO_2 *in vitro*. By the fixation experiments with ^{14}C (see above) we learned that the carbon dioxide is primarily incorporated into a carboxyl group. This is reduced to the carbonyl group of aldehydes, in some organisms even to the stage of an alcohol (like glycerol). It will probably be difficult to simulate the reduction process with the same compounds as used by the living plant. In this case we would need several enzymes and a complicated acceptor: the ribulosebisphosphate. But instead of the organic molecules R—H we may take just H_2.

With a high charge density CO_2 can be electrolytically reduced to formic acid (H—COOH). The next intermediate would be the formaldehyde (H—CHO), the final product methanol CH_3—OH. To optimize the reaction sequence is a real challenge for electrochemists. To come from CO_2 to formic acid we need an electrode material with a high hydrogen overpotential. With this it is, indeed, very easy to obtain a high efficiency. The product itself is a good *reservoir* for hydrogen: we know several catalysts which split it into CO_2 and hydrogen. Much more difficult is the transition from a carboxyl group. This requires special catalysts (which could be found, however). Since the experiments of the Russian chemists BUTLEROV (see above) we know about a photochemical transformation of formaldehyde to a mixture of sugars (*formose*).

It is relatively easy to go from H—CHO to the last product: methanol. Today all the steps are possible under laboratory conditions only. Whether (or when) they become manageable under industrial conditions is a still open question.

At any case the know-how of electrochemists can contribute not only to a better understanding of the natural photosynthetic process but to all efforts to come to its simulation. It certainly requires a closer cooperation between physicists, chemists and biologists to solve the many remaining problems. In some cases we are still looking for suitable methods. But we all should do our best to learn more about the light-induced reaction sequence, to which we owe the existence of life on Earth.

References

[1]   J.R. MAYER, *Die organische Bewegung in ihrem Zusammenhange mit dem Stoffwechsel*, Drechslersche Buckhandlung, Heilbronn (1845).

[2]   J.J. KATZ, J.R. NORRIS and L.L. SHIPMAN, *Brookhaven Symp. Biol.* **28**, 16 (1977).

[3]   S.S. BRODY, *Internat. Kolloq. über schnelle Reaktionen in Lösungen*, Hahnenklee (1969), pp. 366-383.

[4]   Th. FÖRSTER, *Fluoreszenz organischer Verbindungen*, Vandenhoeck and Ruprecht, Göttingen und Zürich (1951).

[5]   R.S. KNOX, in *Bioenergetics of Photosynthesis*, GOVINDJEE (Editor), Academic Press, New York, San Francisco, London (1975), pp. 183-221.

[6]   R. EMERSON and W. ARNOLD, *J. Gen. Physiol.* **15**, 391 (1975).

[7]   G. SCHEIBE, A. SCHÖNTAG and F. KATHEDER, *Naturwissenschaften* **27**, 499 (1939).

[8]   W. ARNOLD and H.K. MACLAY, *Brookhaven Symp. in Biol.* **11**, 1 (1959).

[9]   CIBA Foundation (Editor): *Chlorophyll Organization and Energy Transfer in Photosynthesis*, Excerpta Medica, Amsterdam, Oxford, New York (1979).

[10] R.B. PARK and J. BIGGINS, *Science* **144**, 1009 (1964).

[11] G. BRIEGLEB, *Elektronen-Donator-Acceptor-Komplexe*, Springer-Verlag, Berlin, Göttingen, Heidelberg (1961).

[12] Z. GROMET-ELHANAN, in *Encyclopedia of Plant Physiology*, A. TREBST and M. AVRON (Editors) Springer-Verlag, Berlin, Heidelberg, New York (1977), Vol. **5**, pp. 637-662.

[13] Y. LEMOINE in *Photosynthesis*, G. AKOYUNOGLOU (Editor), Balaban Internat. Sci. Services, Philadelphia, PA (1981), Vol. V, pp. 377-386.

[14] K. TAGAWA and D.I. ARNON, *Nature (London)* **195**, 537 (1962).

[15] B. KOK, *Biochim. Biophys. Acta* **48**, 527 (1961).

[16] F.L. CRANE in *Biological Oxidations*, T.P. SINGER (Editors) *Interscience*, New York (1968), pp. 533-580.

[17] S. KATOH, I. SHIRATORI and A. TAKAMIYA, *J. Biochem.* **51**, 32 (1962).

[18] A. von BAEYER, *Ber. Dtsch. Chem. Ges.* **3**, 63 (1870).

[19] S. RUBEN, M.D. KAMEN and W.Z. HASSID, *J. Am. Chem. Soc.* **62**, 3443 (1940).

[20] D.I. ARNON, M.B. ALLEN and F.R. WHATLEY, *Biochim. Biophys. Acta* **20**, 449 (1956).

[21] B. KOK, in *Handbuch der Pflanzenphysiologie*, W. RUHLAND (Editor) Springer-Verlag, Berlin, Göttingen, Heidelberg (1960), Vol. V, pp. 566-633.

[22] O. WARBURG, *Z. Elektrochem. Angew. Physik. Chem.* **55**, 447 (1951).

[23] R. EMERSON, R. CHALMERS and C. CEDERSTRAND, *Proc. Natl. Acad. Sci.* USA **43**, 133 (1957).

[24] L.N.M. DUYSENS, J. AMESZ and B.M. KAMP, *Nature (London)* **190**, 510 (1961).

[25] A. STANIENDA, *Naturwissenschaften* **50**, 731 (1963).

[26] V.V. KLIMOV, A.V. KLEVANIK, V.A. SHUVALOV and A.A. KRASNOV-SKY in *Photosynthetic Oxygen Evolution*, H. METZNER (Editor), Academic Press, London, New York, San Francisco (1978) pp. 147-155.

[27] A. JAGENDORF and E. URIBE, *Brookhaven Symp. in Biol.* **19**, 215 (1967).

[28] R. WILLSTÄTTER and A. STOLL, *Untersuchungen über die Assimilation der Kohlensäure*, Springer-Verlag, Berlin (1918).

[29] O. WARBURG, *Science* **128**, 68 (1958).

[30] S. RUBEN, M. RANDALL, M. KAMEN and J.L. HYDE, *J. Am. Chem. Soc.* **63**, 877 (1941).

[31] A.P. VINOGRADOV and R.V. TEIS, *Dokl. Akad, Nauk SSSR* **33**, 490 (1941).

[32] T. YOSIDA, N. MORITA, H. TAMIYA, H. NAKAYAMA and H. HUZISIGE, *Acta Phytochim.* **13**, 11 (1942).

[33] C.B. van NIEL, *Arch. Mikrobiol.* **3**, 1 (1931).

[34] A.C. NEISH, *Biochem. J.* **33**, 300 (1939).

[35] H. METZNER, *J. Theor. Biol.* **51**, 201 (1975).

[36] R. HILL, *Nature (London)* **139**, 881 (1937).

[37] J.P. HOARE, *The Electrochemistry of Oxygen*, Interscience Publ., New York (1968).

[38] O. WARBURG and G. KRIPPAHL, *Z. Naturforsch. Teil B*, **13**, 509 (1958).

[39] H. METZNER, *Naturwissenschaften* **53**, 141 (1966).

[40] P. GEORGE in *Oxidases and Related Redox Systems*, T.E. KING and M. MORRISON (Editors), Wiley, New York (1965) Vol. I, pp. 3-36.

[41] J. FRANCK, P. PRINGSHEIM and D.T. LAD, *Arch. Biochem.* **7**, 103 (1945).

[42] P. JOLIOT, G. BARBIERI and R. CHABAUD, *Photochem. Photobiol.* **10**,

309 (1969).

[43] G.M. CHENIAE and I.F. MARTIN, *Biochim. Biophys. Acta* **197**, 219 (1970).

[44] V.M. KUTYRIN, N.M. NAZAROV and K.G. SEMENIUK, *Dokl. Akad. Nauk SSSR* **171**, 215 (1966).

[45] O. WARBURG and G. KRIPPAHL, *Z. Naturforsch. Teil B*, **15**, 788 (1960).

[46] K. WERDAN and H.W. HELDT, *Biochim. Biophys. Acta* **283**, 430 (1972).

[47] J.H.C. SMITH, *Plant Physiol.* **15**, 183 (1940).

[48] K. FISCHER and H. METZNER, *Photobiochem. Photobiophys.* **2**, 133 (1981).

[49] H.-M. BARCHET, *Chemie photographischer Prozesse*, Akademie-Verlag, Berlin (1973).

[50] E. BECQUEREL, *La Lumière, ses Causes et ses Effets*, Paris (1968).

[51] H. METZNER and K. FISCHER, *Progress in Photosynthesis Research II*, Tübingen (1969) pp. 1027-1031.

[52] H. METZNER and K. FISCHER, *Photosynthetica* **8**, 257 (1974).

[53] H. METZNER, K. FISCHER and G. LUPP, *Photosynthetica* **9**, 327 (1975).

[54] H. METZNER and R. GERSTER, *Photosynthetica* **10**, 302 (1976).

[55] G. KORTÜM, *Lehrbuch der Elektrochemie*, Dietrich'sche Verlagsbuchhandlung, Wiesbaden (1948).

[56] H.C. UREY, *J. Chem. Soc.* 562 (1947).

[57] H. METZNER, *Bioelectrochem. Bioenerg.* **3**, 573 (1976).

[58] A.P. VINOGRADOV, V.M. KUTYURIN, M.V. ULUBEKOVA and I.K. ZADOROZHNY, *Dokl. Akad. Nauk SSSR* **134**, 1468 (1960).

[59] H. METZNER, K. FISCHER and O. BAZLEN, *Biochim. Biophys. Acta* **548**, 287 (1979).

[60] H. METZNER, K. FISCHER and O. BAZLEN, in *Photosynthesis,* G. AKOYONOGLOU (Editor), Internat. Sci. Services, Philadelphia, PA (1981), Vol. II, pp. 375-387.

[61] O. REITZ and K.F. BONHOEFFER, *Z. Physik. Chem. A* **172**, 369 (1935).

[62] J. CURRY and S.F. TRELEASE, *Science* **82**, 18 (1935).

[63] H. METZNER, K. FISCHER and O. BAZLEN, in *Stable Isotopes, Anal. Chem. Symp. Ser. 11*, H.L. SCHMIDT, H. FÖRSTEL and K. HEINZINGER (Editors), Elsevier, Amsterdam (1982), pp. 517-527.

[64] E.J. CONSTAM and A.V. HANSEN, *Z. Elektrochem.* **3**, 137 (1896).

[65] P. CUENDET and M. GRÄTZEL, *Photobiochem. Photobiophys.* **2**, 93 (1981).

[66] J. KELLER and R. BACHOFEN, in *Progress in Photosynthesis Research*, H. METZNER (Editor), Vol. II, 1013 (1969).

[67] W. FINKELNBURG, *Einführung in die Atomphysik*, 5./6. Aufl., Springer-Verlag, Berlin, Göttingen, Heidelberg (1958).

[68] F.P. ZSCHEILE and C.L. COMAR, *Bot. Gaz.* **102**, 463 (1941).

[69] G.R. SEELY, in *The Chlorophylls*, L.P. VERNON and G.R. SEELY (Editors), Academic Press, New York, London (1966).

[70] J. MYERS and C.S. FRENCH, *J. Gen. Physiol.* **43**, 723 (1960).

[71] R.K. CLAYTON, *Photobiologie*, Verlag Chemie, Weinheim (1977), Band 2.

[72] W. JUNGE, in *Encyclopedia of Plant Physiology*, N.S., A. TREBST and M. AVRON (Editors), Springer-Verlag, Berlin, Heidelberg, New York (1977), Vol. **5**, pp. 59-93.

[73] B. KOK, in *Harvesting the Sun*, A. SAN PIETRO, F.A. GREER and T.J. ARMY (Editors), Academic Press, New York, London (1967), pp. 29-48.

[74] E.J. DuPRAW, *Cell and Molecular Biology*, Academic Press, New York, London (1968).

[75] H.T. WITT, in *Bioenergetics of Photosynthesis*, GOVINDJEE (Editor), Academic Press, New York, San Francisco, London (1975), pp. 493-554.

[76] H. METZNER, *Biochemie der Pflanzen*, Ferdinand Enke Verlag, Stuttgart (1973).

[77] H. METZNER in *Die Zelle. Struktur und Funktion*, 3 Aufl., H. METZNER (Editor), Wiss. Verlagsgesellschaft, Stuttgart (1980) pp. 313-355.

[78] J.R. BOLTON in *Primary Processes in Photosynthesis*, J. BARBER (Editor), Elsevier/North Holland Publ. Co., Amsterdam (1977), pp. 187-202.

[79] D.E. METZLER, *Biochemistry*, Academic Press, New York, San Francisco, London (1977).

[80] H. METZNER, *Umschau* **75**, 435 (1975).

[81] A. TREBST, *Ann. Rev. Plant Physiol.* **25**, 423 (1974).

[82] B. BOUGES-BOCQUET, *Thesis*, Paris 1974.

[83] K. VEJLBY, *Physiol. Plantarum* **12**, 893 (1959).

[84] J. SINCLAIR, A. SARAI, T. ARNASON and S. GARLAND in *Photosynthetic Oxygen Evolution*, H. METZNER (Editor), Academic Press, London, New York, San Francisco (1978), pp. 295-320.

Further useful readings

J. BARBER (Editor), *Primary Processes of Photosynthesis*, Elsevier Scientific Publ. Co., Amsterdam, New York, Oxford (1977).

J. BARBER (Editor), *Photosynthesis in Relation to Model Systems*, Elsevier Scientific Publ. Co., Amsterdam, New York, Oxford (1979).

R.K. CLAYTON, *Photosynthesis: Physical Mechanisms and Chemical Patterns*, Cambridge Univ. Press, London (1980).

M. GIBBS and E. LATZKO (Editors), *Photosynthesis II. Photosynthetic Carbon Metabolism and Related Processes*, Encyclopedia of Plant Physiology, N.S., in A. PIRSON and M.H. ZIMMERMANN (Editors), Springer-Verlag, Berlin, Heidelberg, New York (1979).

GOVINDJEE (Editor), *Bioenergetics of Photosynthesis*, Academic Press, New York, San Francisco, London (1975).

D.O. HALL and K.K. RAO, *Photosynthesis, Studies in Biology* No 37, Arnold, London (1972).

H. METZNER (Editor), *Photosynthetic Oxygen Evolution*, Academic Press, New York, San Francisco, London (1978).

A. TREBST and M. AVRON (Editors), *Photosynthesis I. Photosynthetic Electron Transport and Photophosphorylation*, in A. PIRSON and M.H. ZIMMERMANN (Editors), *Encyclopedia of Plant Physiology*, N.S., Springer-Verlag, Berlin, Heidelberg, New York (1979), Vol. 5.

C.P. WHITTINGHAM, *The Mechanism of Photosynthesis*, Arnold, London (1974).

ENERGETICS OF BIOLOGICAL REDOX REACTIONS

BRUNO ANDREA MELANDRI

Institute of Botany, University of Bologna
Via Irnerio 42, 40126 Bologna, Italy

Contents

1.   Compartmentation and coupling of biological redox reactions

Biological oxido-reduction reactions, catalyzed by a large variety of enzymes exchanging electrons between different electron donors and acceptors, and various coenzyme molecules, must interact with each other in a ordered and controlled fashion, in such a way to maintain the overall velocity of metabolism in steady state conditions. In a single phase situation, as is the case for soluble redox enzymes in the cytoplasm, the redox coupling between metabolic reactions depends in fact upon the concentrations maintained in the steady state for the different reagents and products, and is also related to the kinetic properties of the various redox enzymes.

The structural organization of multicompartment cells can pose constraints on this kinetic situation: as an example, the coupling of redox reactions through the pool of pyridine nucleotide coenzymes can be discussed. The accumulation on a single reduced coenzyme, like NADH, of reducing equivalents by various dehydrogenases, can offer important functional advantages, since it allows the coupling between a variety of electron donors and different metabolic pathways for the utilization of reducing equivalents for biosynthetic purposes. The regulation of such a network of reactions coupled to each other is, on the other hand, difficult to control through the usual ways of enzyme regulation. The compartmentation of blocks of enzymatic reactions within specialized organelles may contribute to solve these difficulties, since the diffusion barrier of the organelle membrane can be exploited as an additional controlling device. Examples of such a situation are the isolation of the oxidative metabolism of the KREBS cycle within the mitochondrial matrix, or of the CO_2 reductive cycle in the chloroplast stroma. In these cases the redox coupling with the cytoplasmic reaction must be of course maintained, but this interaction is mediated, and regulated by the transport processes across the periferal membranes of the organelles, not directly of the oxidoreduction coenzymes but rather of metabolite with a suitable degree of oxidoreduction. An example of this type of interactions is shown in Fig. 1 for higher plant chloroplasts [1]; in this case the export of the reducing power produced originally as NADPH by photosynthesis is mediated either by the operation of the malate-oxaloacetate shuttle, or by the 3-hydroxyacetonphosphate-3 phosphoglycerate transporter. Since in the latter case the going out of the NADPH equivalent is compulsorily coupled to that of one molecule of ATP, a combination of both transport systems allows

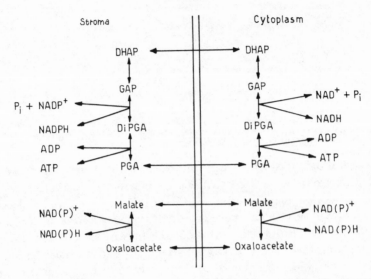

Fig. 1. — Schematic representation of the metabolite flows catalyzed by the DHAP and malate transporters across the chloroplast outer envelope.

the necessary flexibility to the stroma-cytoplasm interaction as far as the reducing power and the energy charge are concerned.

In other situations a more strict and specialized redox coupling of the different oxidoreductive steps can be convenient. This strict coupling is accomplished through a supramolecular organization of the enzymes, such as the formation of multienzyme complexes or the integration of the redox carrier in membrane structures. As examples of these two situations the structure of the pyruvate dehydrogenase complex from E. coli and of the microsomal NADH-cytochrome b_5 systems of liver will be described.

The mechanism of the pyruvic dehydrogenase reaction, catalyzing the conversion of pyruvate to acetyl-CoA, is depicted in Fig. 2: the reaction goes through three different steps, each catalyzed by a different enzyme which is part of the multienzyme complex [2]. The first is a decarboxylation step which involves a TPP-dependent enzyme, pyruvate dehydrogenase; the second step, catalyzed by the dihydroxylipoyltransacetilase, transfers the acetyl residue from TPP to CoA-SH; the third, mediated by the dihydrolipoyldehydrogenase, a FAD-containing enzyme, oxidizes the reduced lipoyl prostetic group with NAD^+. The three enzymes, purified separately from E. coli, form spontaneously a multienzyme complex of 72 subunits, formed by 24 copies of the single polipeptide [3]: for the formation of this complex the transacetylase is always necessary, while the absence of each one of the other two enzymes does not prevent the formation of incomplete binary complexes. It is

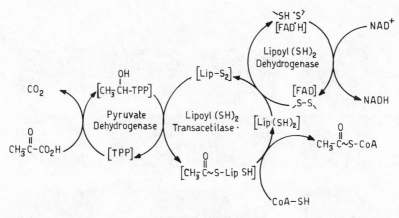

Fig. 2. — The mechanisms of the pyruvate dehydrogenase reaction catalyzed by the pyruvate dehydrogenase multienzyme complex.

therefore believed that the transacetilase not only plays a central role in catalysis, but also forms the structural core of the complex. It has been also proposed that the lipoic acid residue, attached to a lysyl residue and forming a structure 1.4 nm (14 Å) long, can transfer, with an oscillatory movement, the acetaldehyde residue from TPP to a reactive disulfide of the flavoprotein [4].

Electron micrograph images of the three different enzymes and the complex have been obtained; the schematic drawings of Fig. 3 show the high order-degree obtained in this supramolecular structure [5]. In this tight packing of the three kinds of subunits the highly favourable position of the three enzymic components causes a strict coupling of the different oxidoreductive steps of the overall reaction, minimizing the energy losses and unspecific interactions that are likely to occur in a soluble enzyme chains. Thus the structural organization results in a compartmentation of the redox events and in an improvement of the specificity and efficiency of the redox coupling.

In the liver microsomal system the strict coupling of the NAD-cytochrome b_5 oxidoreductase is obtained by binding the enzymes at the membrane surface. The two known components of this redox chain, a flavoprotein NADH-cytochrome b_5 reductase and cytochrome b_5 can be obtained from microsomal preparation in two forms (cfr. Table 1): a water soluble preparation, obtained by the protease treatment of the membranes, and a more hydrophobic one prepared by micellization of the membrane with detergents [6]. For both proteins the two preparations differ for the presence (in the detergent preparation) or absence (in the water

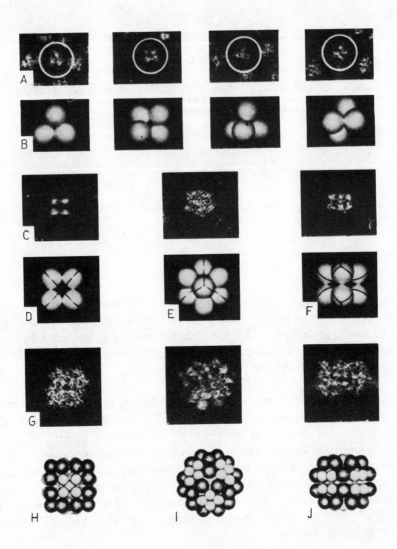

Fig. 3. — Electron microscopic images and interpretative models of the pyruvate dehydrogenase complex and its subunits. (A,B) isolated pyruvate dehydrogenase (x 700 000) at various orientations; (C,D,E,F) images (x 350 000) and models of transacetylase, consisting of a cubic structure formed by eight three-subunits spheres (total 24 subunits); (H,G,J) various orientations of the whole complex, in which 24 pyruvate dehydrogenases and 24 lipoyl dehydrogenases are distributed along the edges of the cubic structure (C,D) (from Ref. 49).

Table 1. — Comparison of the amino acid composition of NADH-cytochrome b_5 and cytochrome b_5 reductase from rat liver, prepared by detergent treatment or snake venom digestion of microsomal membrane (data from SPATZ and STRITT-MATTER [6, 9]).

Amino acid	NADH-cytochrome b_5 reductase			Cytochrome b_5		
	Detergent Extracted A	Protease Extracted B	A-B	Detergent Extracted A	Protease Extracted B	A-B
Cysteic acid	7	6	1	0	0	0
Aspartic acid	34	28	6	16	10	6
Threonine	16	14	2	10	7	3
Serine	19	13	6	10	7	3
Glutamic acid	38	29	9	15	14	1
Proline	33	29	4	5	3	2
Glycine	27	20	7	7	6	1
Alanine	21	15	6	10	5	5
Valine	28	16	12	7	4	3
Methionine	10	8	2	3	1	2
Isoleucine	24	19	5	8	4	4
Leucine	40	27	13	15	9	6
Tyrosine	13	8	5	5	3	2
Phenylalanine	18	13	5	4	3	1
Tryptophan	6	2	4	4	1	3
Lysine	25	20	5	11	10	1
Histidine	12	9	3	7	7	0
Arginine	20	16	4	4	3	1
Total residues	391	292	99	141	97	44
Molecular weight	44 185	32 840	11 340	16 072	11 079	4 993

soluble form) of a polipeptide segment, but are virtually identical in their catalytic properties (cfr. Table 1). Since the presence of the cleavable peptides is essential for the attachment of the proteins to the microsomal membrane [7], it has been proposed that they form a hydrophobic structure that anchors the proteins, *per se* extrinsic to the membrane structure, to the membrane lipids. The kinetic analysis of the overall reaction has demonstrated that the flavoprotein-cytochrome b_5 interaction is limited by the lateral diffusion of these proteins on the membrane surface [8]. The association of these systems to a membrane results merely in a rather mild kinetic constraint in the interaction of these enzymes, and can therefore be considered as a sort of compartmentation.

The components of the oxidoreductive chains in energy transducing membranes of respiratory, photosynthetic or chemosynthetic systems are subjected to much stricter kinetic controls. In this case the electron transport carriers of the chain (flavoproteins, cytochromes and iron-sulphur proteins) are generally intrinsic component of the membrane structure and are organized in intramembrane multienzyme complexes, in which the oxidoreductive function is coupled to the capacity of conserving and transducing the free energy of the oxidoreductive reactions. In such organized structures the kinetic controls are more rigorous and result both from the interaction of the electron carrier within the multienzyme complexes, and from the compartmentation of these complexes *within* a single membrane. Given their predominant function in cellular bioenergetics, these electron transport chains are of central interest in bioelectrochemistry and must be studied experimentally both from a thermodynamic and kinetic viewpoint. The following sections will be dealing therefore with such systems.

2. Thermodynamic characterization of enzymatic redox systems in biological membranes

The evaluation *in situ* of the thermodynamic properties of the redox systems present in energy transducing membranes (specifically of cytochromes, iron sulphur proteins, and flavoproteins) is of utmost importance for the study of the bioenergetics of these systems and for the experimental test of their energy transducing mechanisms. The redox catalysts present in respiratory or photosynthetic membranes of eucariotes or procariotes are believed to be coupled to proton translocating pumps [10] and to be able therefore to perform electrochemical work during their redox cycles; they are moreover coupled to endoergonic group transfer reactions in the processes of oxidative or photosynthetic phosphorylation [11].

The most extensively studies of these redox systems is the mammalian mitochondrial respiratory chain, which will be amply discussed in another chapter of this book. Therefore the examples that will be used in this chapter are mainly taken from the photosynthetic system of facultative photosynthetic bacteria with which the present author is more familiar [12].

Membrane redox systems are generally formed by intrinsic membrane protein hardly accessible to experimentation since they are deeply buried in the membrane lipid core. The more direct way to evaluate their thermodynamic properties is the measurement of their midpoint potential (U_m') *in situ*. For this experimental approach the redox state of a single catalyst must be evaluated as a function of the ambient redox potential, under conditions in which an oxidoreductive equilibrium is established [13].

Different techniques can be utilized for the evaluation of the redox state of the protein, which use specific spectroscopic properties of the different prostetic groups; thus for cytochromes, dual-wavelength spectrophotometry in the α-band absorption region has been largely utilized, as well as *e.p.r.* spectroscopy for iron-sulfur centers and, to a smaller extent, fluorometry for flavoproteins. These techniques, under certain specific conditions, allow the study of single chromophores, or at least a single class of chromophores, without large interferences from others, as will be discussed in the following sections.

The redox equilibrium between the chromophore under study and the measuring electrode (generally a Pt electrode *versus* a saturated KCl calomel reference electrode) must be carefully established; this is generally accomplished with the use of a mixture of redox mediators, whose midpoint potential should be chosen in order to cover the whole experimental U_h range, and which should be sufficiently hydrophobic to interact with prostetic groups present within the membrane structure [13]. In addition they should not interfere spectrally with the chromophores studied. A list of the mediators most commonly used in membrane potentiometry is given in Table 2. Most of these mediators are easily autooxidizable and therefore the measurements should be performed under a nitrogen or argon atmosphere.

Table 2. — Properties of the most commonly used mediators in redox potentiometry. Data from Ref. 13.

Redox mediator	$U^{\circ}_{m\,7}$ (mV)	n value
2, 3, 5, 6 tetramethyl phenilenediamine (DAD)	260	2
N, N, N′, N′ tetramethyl phenilenediamine (TMPD)	260	1
N-methyl phenazonium methosulphate (PMS)	80	2
N-ethyl phenazonium ethosulphate (PES)	55	2
N-methyl-1-hydroxyphenazonium methosulphate (pyocianine)	− 34	2
2-hydroxyl-1,4 naphtoquinone	− 145	2
Antraquinone-2-sulphonate	− 225	2
N-N′-dibenzyl-4,4 bipyridinium HCl (Benzyl viologen)	− 311	1
N-N′-dimethyl-44 bipyridinium HCl (methyl viologen)	− 430	1

When a spectroscopic property of a single chromophore, *e.g.* the differential absorbance of the α-band of a cytochrome at a specific wavelength pair, is linear with the extent of reduction, the redox state can be followed directly as a function of the ambient U_h (at equilibrium through the mediators mixtures with the enzyme studied) while the U_h is gradually and reversibly altered with small additions of a reductant (usually with ascorbate or sodium dithionite) or an oxidant, $K_3[Fe(CN)_6]$. From the experimental data a nernstian plot can be drawn, generally in a semilogarithmic form, which allows the evaluation of the midpoint potential (U'_m) and of the n value (see Fig. 4), the number of electrons involved in the redox cycle.

When this kind of measurements are performed at different pH's and if the experimental values of U'_m are found to be pH-dependent, the acid-base properties of the redox species can also be evaluated. A general relation linking the apparent U'_m, the true U'^o of a given redox couple and the pK values of the oxidized and reduced species according to the following equilibria, can be obtained:

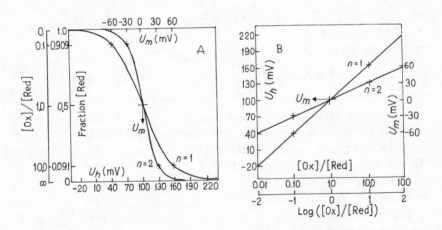

Fig. 4. — NERNST plot for two redox couples with $n = 1$ or 2 respectively. A) linear plot of the Ox/Red ratio *versus* ambient potential. B) semilog plot of the data of A. The U_m value is determined from the interpolation at Ox/Red = 1. The slope of the semilog plot is 60 mV and 30 mV per Ox/Red decade for $n = 1$ or 2 respectively. (modified from Ref. 48).

$$
\begin{array}{ccc}
\text{Red·H} & \xrightarrow{\qquad U_m \ (\text{acid})\qquad} & \text{Ox}^+ \text{H} + \text{e}^- \\
\Big\updownarrow K_{\text{Red}} & & \Big\updownarrow K_{\text{Ox}} \\
\text{Red} + \text{H}^+ & \xrightarrow{\qquad U_m \ (\text{bases})\qquad} & \text{Ox} + \text{e}^- + \text{H}^+
\end{array}
$$

This relation for a monovalent, monoprotonated redox species ($n = 1$), is (at 25 °C):

$$
U_h = U_m + 0.06 \log \frac{[\text{Ox}^+\text{H}]}{[\text{Red H}]} + 0.06 \log \frac{1 + (K_{\text{Ox}}/[\text{H}^+])}{1 + (K_{\text{Red}}/[\text{H}^+])} \tag{1}
$$

The ideal titration of a single species at different pH's is shown in Fig. 5, and the determination of pK_{Ox} and pK_{Red} from the experimental U_m' is illustrated. This kind of behavior, which is very often observed during titrations of redox components of the respiratory or photosynthetic chain, offer an experimental basis for the identification of protonable catalysts, possibly involved in the proton translocation mechanism [12].

When more than one redox component is monitored at a given wavelength pair, as shown *e.g.* in Fig. 6 for the cytochromes of *b* type in the respiratory membrane from *Rhodopseudomonas capsulata* [14], a composite pattern of reduction

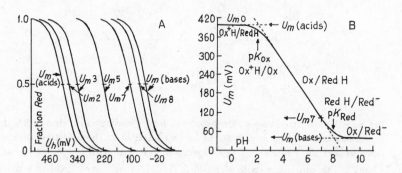

Fig. 5. – The dependence of U_m from pH for a redox couple with pK(acids) = 2 and pK(bases) = 8. The apparent U_m decreases linearly with pH with a slope of 60 mV per pH unit (1 H$^+$ and 1 e$^-$ per molecule) except below pH = 2 or above pH = 8 where it become pH insensitive. (modified from Ref. 48).

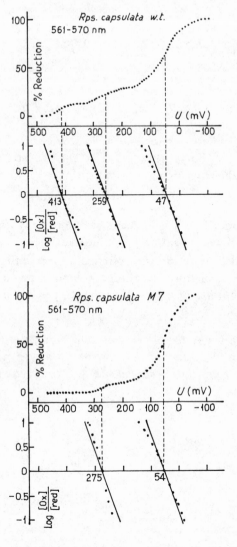

Fig. 6. — Redox titrations of cytochromes of *b* type in chromatophores of *Rhodo-pseudomonas capsulata*, strains St. Louis (w.t.) and M 7. The wild type titration demonstrates the existence of three species of cytochromes *b* ($U_m = 413, 259$ and 47 mV). The experiment with strain M 7, lacking cytochrome *c* oxidase activity, shows the absence of cytochrome b_{413}, a plausible candidate as cytochrome oxidase. (from Ref. 14).

is observed; in this case, generally the different components overlap during the redox titration so that in a given U_h span the observed extent of reduction contains contributions of different redox species. The most accurate method for best fitting such experimental data is the digital computation of the contribution of the different species, with computer programs which optimize the values for the different U_m' and for the concentration of the components contributing to the overall redox spectral response [15].

An example of a biological application of redox potentiometry is also shown in Fig. 6; the data expressed in the lower diagram represent titration of b cytochromes in a mutant strain of *Rps. capsulata*, strain M7, lesioned in the cytochrome c_2 oxidase activity. It is clear that in this preparation one cytochrome b, with U_m' at 413 mV, was absent; this evidence was therefore very strong support to the idea that cytochrome b 413 was the respiratory component catalysing cytochrome oxidase, a concept substantiated by other subsequent findings.

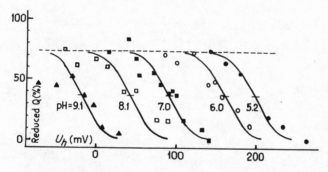

Fig. 7. — Redox state of extracted ubiquinone-10 of *Rhodopseudomonas sphaeroides* at different ambient pH. The data, obtained extracting UQ at controlled U_h and pH demonstrate the dependence of the apparent U_m of this carrier from pH. (from Ref. 16).

Another example of redox titration, always in photosynthetic bacteria, is [concerned with] the thermodynamic characterization of ubiquinone of *Rhodopseudomonas sphaeroides*. In this instance, a direct spectroscopic observation of the redox state of ubiquinone in the membrane is not feasible, since ubiquinones are UV absorbers which overlap heavily with the spectra of membrane pigments and of the aromatic residues of aminoacids. The data, depicted in Fig. 7 were therefore obtained by extracting ubiquinones with organic solvents under controlled redox conditions and evaluating the redox state in the extract [16]; the scattering of the experimental points is accounted for by the inaccuracy of such a two step technique. The evaluation of the apparent U_m at different pHs demonstrates that midpoint potential depends on the proton activity, with a dependence of 60 mV per pH unit. Thus as a whole the potentiometric data indicates that the bulk of the ubiquinone, present in great abundance in photosynthetic bacterial membranes, behaves as a 2 e⁻ and 2 H⁺ carrier from pH 5.0 to pH 9.0 [13].

More indirect kinetic approaches can be utilized for obtaining thermodynamic information on electron carriers spectroscopically silent. For example the experiment shown in Fig. 8 illustrates the kinetics of reduction of cytochrome c, following flash-photooxidation as a function of the ambient redox potential. It is evident that there is a marked acceleration of the rate of reduction for $U_h \leqslant 200$, indicating the presence of a reductant of cytochrome c in this range of potentials. This carrier titrates with an $U_{m\,7} = 150$ mV. It was subsequently demonstrated to be one or a few molecules of ubiquinone, whose thermodynamic properties are significantly different from those of the large omogeneous pool discussed above [17, 18]. In this case therefore, information concerning an electron carrier has been drawn from the kinetic behavior of another carrier whose redox state can be conveniently monitored spectroscopically.

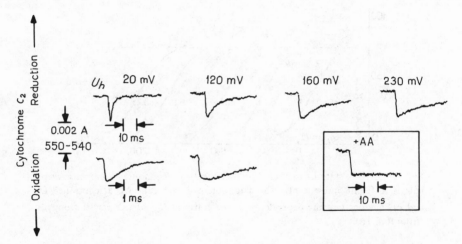

Fig. 8. — Reduction kinetics of photooxidized cytochrome c_2 in chromatophores of *Rhodopseudomonas sphaeroides* at different ambient redox potentials. The reduction rate, which is inhibited completely by the antibiotic antimycin A (panel), is strongly accelerated when the U_h is lowered from 200 to 20 mV. The U_m of this effect is 150 mV at pH = 7.0. This experiment demonstrates the existence of a reductant of cytochrome c_2 with an apparent U_m of this value (cfr. text) (modified from Ref. 17).

From similar approaches many redox carriers can be identified in biological membranes and their thermodynamic and proton accepting properties can be determined. This information can be complemented with kinetic information on the rate constants of the different steps of electron transport. Eventually these studies

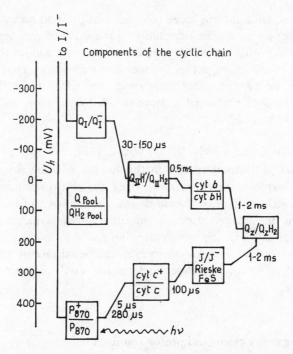

Fig. 9. — Diagrammatic representation of the electron transport chain of *Rhodo-pseudomonas capsulata* or *Rps. sphaeroides*. The number indicate the half time of the different redox steps (for cytochrome *c* the kinetics is biphasic); the boxes cover 60 or 30 mV spans according to the *n* value of the different carriers (modified from Ref. 12).

lead to model of electron transport chains; one example, always for facultative photosynthetic bacteria, is presented in Fig. 9. From a model of this sort further studies on the mechanism of proton translocation and of energy transduction by these systems can be developed.

Before discussing these aspects, however, some points of the uncertainty of redox potentiometry should be considered. This technical approach has also been used in membrane during active electron transport or ATP hydrolysis; under these conditions a different electrostatic potential is developed at the two sides of the membrane, with a polarity depending on the polarity of the membrane vesicles used (negative inside for mitochondria and whole bacterial cells, positive inside for chloroplast, chromatophores and submitochondrial particles). However since only the outside compartment is at equilibrium with the measuring electrodes, the measured apparent U'_m of a chromophore facing the inaccessible compartment will be displaced from its true value (corresponding to that measured in an unpolarized membrane) by an extent depending on the value of the membrane potential, on the potential profile within the membrane dielectric and on the actual position of the chromophore in respect to the membrane thickness. Thus several unknown parameters are introduced under these conditions which make the univocal interpretation of the data problematic. More detailed discussions on this point can be found in Ref. 19 and 20.

As discussed extensively by D. WALZ [20], electrostatic phenomena can influence redox potentiometry also in the absence of a membrane potential. This influence is exerted at the interfaces, where the electrostatic interactions of the redox species with the ionic diffusion double layer and with fixed charges, possibly present on the membrane, affect the apparent thermodynamic properties of the electron carriers and make the redox titration markedly non-nernstian. Under such conditions the complex pattern of titration can be mistakenly interpreted as evidence of several redox species, while the distortion is only due to electrostatic effects at the interface.

3.   Coupling redox events and proton translocation

According to the chemiosmotic theory, in respiration and photosynthesis the redox reactions catalyzed by the electron transport carriers present within the membranes are coupled to the electrogenic translocation of protons across the membrane [10]. In order to obtain thermodynamic information on this process on the basis of the formalism of non-equilibrium thermodynamics [21], the fluxes and forces involved in this process must be evaluated. These fluxes are the rates of redox reactions (whose evaluation in linear respiratory and photosynthetic systems is generally easily achieved) while the extent of protonic fluxes is generally evaluated with H^+ - reversible electrodes or pH indicator dyes (see below). The forces coniugated to these flows are the free energy changes of the redox reactions, which can be evaluated from the midpoint potentials extensively discussed in the previous paragraph and from the redox state of the electron donors and acceptor during steady state, as determined by spectroscopic techniques. The second force of the process, the difference in electrochemical potential of protons across the membranes, which measures the amount of electrochemical work performed for the translocation of one mole of protons, is generally more difficult to evaluate, and will be briefly discussed here.

Formally, the difference of electrochemical potential of protons is formed by two components, an activity difference (Δ pH) and a difference in electrostatic potential $\Delta\psi$, so that the overall $\Delta\tilde{\mu}_{H^+}$ can be expressed as:

$$\Delta\tilde{\mu}_{H^+} = F\,\Delta\psi - 2.3\,RT\,\Delta\,pH \tag{2}$$

Different techniques have been proposed, and will be described here, for the separate evaluation of the membrane potential and of ΔpH.

3.1. Evaluation of ΔpH

The evaluation of the activity difference of protons in biological micro-vesicular systems is always obtained indirectly, given the general inaccessibility of the inner compartment to experimental determinations. This evaluation is based on the equilibrium distribution of weak amines or weak acid across the biomem-brane according to the following scheme (Fig. 10).

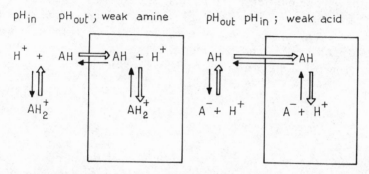

Fig. 10. — Schematic representation of the distribution ratio of a weak amine or acid in a two compartment system, as a consequence of an equilibrium distribution of the uncharged species and of the protonation equilibria in the two water phase.

In the scheme, the thick arrows indicate the displacement of diffusion and protonation equilibria due to the asymmetry in proton activity in the two compart-ments, so that for a properly choosen marker (amine or acid as indicated in the scheme) a net accumulation of the marker ion is achieved. Formal treatment of the three equilibria demonstrates that, for pH_{in} and pH_{out} distant enough from the pK_a of the marker, the activity ratio in the two compartments is only a function of ΔpH [22]. A necessary condition for such a distribution is the fast equilibration of the undissociated species across the membrane and the negligible permeation of the charged species (A^- or AH_2^+) [22].

The evaluation of the distribution ratio is generally achieved by measuring the accumulation of isotope labelled marker ions and using fast separation tech-niques such as filtration or centrifugation [23]. Alternative methods for the evaluation of the external concentration of the marker are flow dialysis [24], reversible electrodes (these restricted only to the amine NH_3 [25]) or fluorescence (for fluorescing amines like 9amino-acridine [26]); in these cases the accumulated species is evaluated indirectly by the mass balance of the total ion disappearing from the outer phase. With any method, the calculation of the internal marker concentration requires the evaluation of the volume of the inner space, generally obtained with isotopic techniques.

A point of uncertainty in this method is offered by the exstimation of the activity coefficient of the marker ion in the internal compartment; any interaction (binding) of the marker with components of the biological systems, such as electrostatic or hydrophobic interactions, will generally contribute to a decrease in the activity of the ion as compared to its concentration, expecially in the inner lumen, in which the ion accumulates several tenfolds at least. This deviation from an ideal thermodynamic behavior of the ion marker should be accurately evaluated for an exact determination of ΔpH; since, however, the inner compartment is inaccessible to direct experimentation and the internal concentration is evaluated indirectly, this point is generally neglected and the activity coefficient is taken as one.

3.2. Evaluation of $\Delta\psi$

The most general method for the evaluation of the membrane potential is again an ion distribution method which rests on the assumption of an equilibration of a marker ion; this approach follows the very definition of electrochemical potential, which is given on the basis of an equilibrium distribution of ion activity balancing the electrostatic potential difference between two phases [23].

The equilibrium equation of a tracer ion between two phases (I and II) at different electrostatic potential is:

$$\Delta\psi = \psi_I - \psi_{II} = \frac{RT}{n\,F} \ln \frac{a_{II}}{a_I} \tag{3}$$

where n is the electric charge of the ion including its sign. Since it is always necessary, for practical experimental reasons, to follow an accumulation of the tracer in the inner compartment, anions must be used for positive inside vesicles and cations for negative inside organelles. The cation chosen are generally K^+ (or $^{86}Rb^+$) in the presence of the carrier antibiotic valinomycin; more difficult is the choice of a suitable anion, since SCN^-, usually utilized [27], is not always freely permeating the natural membrane, and more hydrophobic synthetic ions (like tetraphenylborate [TPB⁻]) bind strongly to the membrane and exert a dramatic uncoupling effect [28].

The technique used for the determination of the distribution ratio of the traces are similar to those utilized for amine or acid distribution, normally rapid centrifugation or filtration of the organelles, flow dialysis [27] or reversible electrodes (for K⁺ or TPB⁻ uptake).

An alternative method for the evaluation of $\Delta\psi$, largely utilized in photosynthetic membrane, is based on the electrochromic phenomena of photosynthetic pigments [29]. This method is particularly useful with bacterial chromatophore for which an accurate calibration of the electrochromic response *versus* the membrane potential can be obtained [30].

Reversible red shifts of the absorption spectrum of the endogenous carotenoid can be readily observed upon illumination of chromatophore suspensions either with continuous light or with short (ns or µs long) flashes of actinic light. These shifts can also be observed in whole cells induced either by light or by oxygenation. On the basis of their response to uncoupler or to inhibitors of electron transport or of phosphorylation, these spectral signals were very soon recognized to be related to the high energy state of the membrane and are nowdays believed to be associated to the onset of an electric field across the chromatophore membrane. Basic for this idea, originally put forward for chloroplasts by JUNGE and WITT [31], were the experiments by JACKSON and CROFTS [30], which demonstrated that shifts, spectrally indistinguishable from the light-induced ones, could be induced in the dark by diffusion potentials of ions (*e.g.* by K⁺ or H⁺ fluxes across membrane treated with valinomycin or FCCP respectively). This approach is today utilized for the calibration of the signal in terms of membrane potential (see below).

The molecular mechanism of the spectral response of carotenoids to an external electric field is currently interpreted as an intrinsic electrochromic effect of these pigments. A simple model for electrochromisms [32] (*i.e.* of the electrostatic influence on the energy states of the chromophores due to the differential dipolar character of the excited and ground states) predicts a shift of the maxima of the absorption bands of the spectra according to the equation:

$$\Delta\lambda = \frac{\lambda^2}{h\,c} \left(\Delta\mu\,E\,\cos\theta + \frac{1}{2}\,\Delta\alpha\,E^2\,\cos^2\phi \right) \tag{4}$$

where h and c are the PLANCK constant and the light velocity, E the applied external electric field, $\Delta\mu$ the change in the permanent dipole moment of the excited *versus* the ground state and $\Delta\alpha$ the change in polarizability of the chromophore. θ and ϕ are the angles between the electric field and $\Delta\mu$ or $\Delta\alpha$ respectively. Thus a linear relation between E (and therefore with $\Delta\psi$ for a constant membrane thickness) and $\Delta\lambda$ has to be expected for permanently polarized chromophores and a quadratic response for polarizable chromophores. Both these responses have been demonstrated for intrinsic carotenoid of chromatophores, but, as we will see in the following, the linear response is largely predominant. Moreover this linear response can be positive or negative (*i.e.* towards the red or the blue region of the spectrum) according to the value of θ, *i.e.* the sideness of the membrane *versus* E; this property has also been experimentally verified [33].

Experimentally, the extent of the shift is generally evaluated as a relative change in absorbance at a fixed wavelength pair, since it can be shown in a first approximation, that

$$\Delta A(\lambda) = -c\, \frac{\delta\epsilon}{\delta\lambda} \Delta\lambda \tag{5}$$

where $\Delta A(\lambda)$ is the change in absorbance at the wavelength λ, c is the concentration of the chromophore and $\delta\epsilon/\delta\lambda$ is the first derivative of the extinction coefficient *versus* wavelength at the wavelength λ. Therefore for small shifts, like those induced by light in chromatophores, the differential spectrum plus or minus and external electric field approximates the first derivative of the absorption spectrum of the responsive chromophore [29].

Calibration of the shift in terms of membrane potential relies on the onset of ion diffusion potentials of known intensity across the membrane and measurements of the resulting spectral effects. If, for example, valinomycin-treated chromatophores are subjected to a sudden addition of KCl, the rapid permeation of K^+ will rapidly charge the membrane electrically, according to its electric capacitance C, and very quickly an electrochemical equilibrium of K^+ at a given $\Delta\psi$ will be established. The formal treatment of this situation yields the following relation between the induced $\Delta\psi$ and the amount of KCl added in the pulse [34]:

$$\Delta\psi = \frac{RT}{F}\, \ln \frac{[K^+]_0 + [K^+]_{add}}{[K^+]_0 + (C\Delta\psi/F\, V_i)} \tag{6}$$

where $[K^+]_{add}$ is the concentration of K^+ added, $[K^+]_0$ is the initial concentration of endogenous potassium present in the chromatophores before the addition and

V_i is the internal volume of the vesicles. The graphic presentation of equation (3), given in Fig. 11, shows that $\Delta\psi$ becomes asimptotically linear with log. $[K^+]_{add}$, with a slope that approaches the NERNST dependence (59 mV per concentration decade at 25 °C).

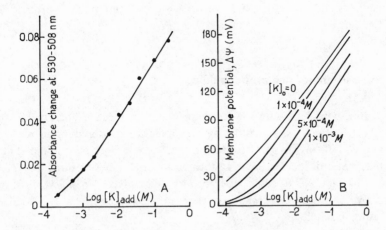

Fig. 11. — Experimental determination of the carotenoid electrochromic response to K^+ diffusion potentials (A) and theoretical dependence of $\Delta\psi$ as a function of $[KCl]_{add}$ at different $[K^+]_0$. The theoretical curves were calculated assuming a capacitance of 1 $\mu F\ cm^{-2}$, an average chromatophore surphace of $10^{-10}\ cm^2$ and an average bacteriochlorophyll content of 5 000 molecule per chromatophore. (from Ref. 34).

The main advantage of the technique of the shift is kinetic, since the electronic response of the pigments is practically instantaneous and information on the electrogenic events of electron transport can therefore be obtained in real time. A similar approach has also been used utilizing exogenous dyes (oxanol, merocyanine) [35, 36]; the response of these probes is however generally much slower than that of endogenous pigments since the spectral change does not always depend only on electrochromic phenomena, but also on the redistribution of the dye induced by the potential difference either between the two compartments or at the interfaces.

In general when one compares the values of $\Delta\psi$ evaluated in the same system by ion distribution and by electrochromic techniques, a significant difference is found. This discrepancy does not necessarily mean a fault of one of the two techniques, but may reflect true differences in the local potential measured by the two approaches. As mentioned in a previous section in fact, fixed electric charges in the membrane can alter the voltage profile at the membrane water interfaces [20], giving rise to ion diffuse double-layer and to localized voltage differences at the interface. The two abovementioned techniques will systematically sense this difference since ion distribution is always evaluated on the whole inner volume of the vesicle in which the bulk water volume is often prevailing, while carotenoids sense the potential from within the membrane and are therefore very sensitive to interface differences. For a more extensive discussion of this point the reader should consult Ref. 12 and 37.

4. Charge separation and proton binding

A current model for redox-driven proton translocation, originally proposed by MITCHELL, suggests this process is coupled to oxidoreduction reactions by *proton loops* [10]; in this model electrons are separated across the membrane and protonated reduced species are translocated in the opposite direction, so that a net transmembrane electrogenic H^+-transport takes place. These events should be catalyzed by multienzymes complexes plugged across the membrane lipid bilayer and operating in series.

Therefore, in order to obtain experimental proof for this model, separated evidence should be obtained for charge separation catalyzed by membrane redox complexes on one hand, and on the other, for protolytic events on the two opposite faces of the membrane coupled to oxidoreductions.

These electrogenic or proton-conducting phenomena should thermodynamically and kinetically match different steps of the redox reactions if they are to be considered in the framework of the loop model.

Evidence for a transmembrane separation of charges by a redox enzyme complexes has been obtained, *e.g.*, with the reaction centers of photosynthetic bacteria and of higher plant chloroplasts. Initial support for this idea came from the observation that electrochromic signals were kinetically matching the primary photochemical reactions indicating a vectorial arrangement of the photosynthetic oxidoreduction [38]. Refined analysis of these phenomena demonstrated moreover that in chromatophores the onset of the electrochromic signal was biphasic, presenting a phase coincident with the primary charge separation within the reaction center and a second one associated with the transfer of one electron from cytochrome c_2 to $[BChl]_2^{+\cdot}$ [39]. From these observations, structural concepts on the arrangement of the reaction center across the membrane were inferred [39]. Similarly, in chloroplasts the contribution to the field formation by the primary events of photosystem I and II could be experimentally separated [40].

Some of the deductions drawn from electrochromism on bacterial reaction centers could be fully confirmed when isolated preparations of this complex were successfully incorporated into black lipids membranes [41]. Stable photocurrents and photovoltage could be induced in this system by continuous illumination, with an action spectrum corresponding to the absorption spectrum of the center pigments, demonstrating the transmembrane arrangement of the oxidoreduction process. For a successful reconstitution, an excess of secondary electron acceptor and donor, respectively ubiquinone-10 and cytochrome c, had to be added asymmetrically to the artificial membrane in order to act respectively as an electron sink and source, and to select artificially only the population of centers properly oriented across the membrane [41].

Also in respiratory systems several observations document the charge separating capacity of the oxidoreduction complexes; the lack of availability of fast techniques to monitor the membrane potential in the absence of endogenous pigments and the necessity of inducing the redox reactions with additions of substrates has prevented so far the advancement of these studies on a refined kinetic basis as compared to the photosynthetic counterpart.

According to the loop model, the protolytic events of proton translocation should be coupled to the redox reactions according to the general scheme presented previously. A first consequence of these mechanism therefore should be the dependency of the apparent midpoint potential upon pH for all those components involved in proton translocation.

Indeed, pH dependency of U_m' has been demonstrated for several components of the respiratory or photosynthetic electron transport, and in particular for several cytochromes of the b type [42], for iron sulfur proteins [43] and quinone coenzymes [16]. These components should therefore be considered to be likely candidates as a catalyzer of proton translocation.

The thermodynamics and kinetics of fast proton binding reactions, occurring upon reduction of membrane electron transport components has been particularly studied in a photosynthetic system in which light-induced electron transport is easily controlled. The proton disappearance has been monitored spectroscopically on the outer face of the membrane by adding pH indicator dyes, chosen among the more hydrophilic to minimize the interactions with the membrane lipid phase (*i.e.* anionic pH indicator like cresol-red or phenol-red) [44].

The data presented in Fig. 12 are an example of such experiments performed on bacterial chromatophores; the binding of 2 H^+ per electron going through the photosynthetic cycle has been demonstrated by flash spectroscopy using cresol red, with one of the two protons being inhibited by the antibiotic antimycin A, and therefore being considered to bind to some component of the ubiquinol-cyto-chrome c oxidoreductase. Similarly, in chloroplast, proton-binding upon plasto-quinone photoreduction has been proved.

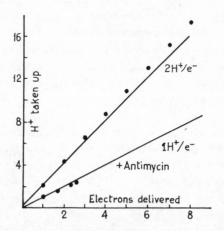

Fig. 12. – Multiple single-turnover-flash-induced binding of protons by chromato-phores of *Rhodopseudomonas sphaeroides* as detected with the pH indicator cresol red at pH = 6.0. The photosynthetic reaction was activated by repetitive flashes fire 40 ms apart. The number of electron driven through the redox chain was deter-mined spectroscopically from the photooxidation signal of the reaction center bacteriochlorophyll doublet. (modified from Ref. 44).

Several experimental discrepancies have however been revealed by such tech-niques. For example in broken chloroplasts there is no kinetic coincidence between the rate of plastoquinone reduction ($t_{1/2}$ = 2 ms) and H^+- that of binding ($t_{1/2}$ = 60 ms), unless the membrane are harshly treated with sand grinding or with detergents [45]. In chromatophores no clear correlation has been obtained between the func-tional pK's evaluates from proton binding experiments and pK values obtained from equilibrium U_m' determination [46]. Thus in chromatophore it is still practically impossible to relate precisely any electron transport component of the photosyn-thetic cycle to protolytic reactions according to a proton-loop scheme.

The above mentioned studies performed with hydrophilic pH indicator can report only on events taking place on the outer face of the membrane, on which, in photosynthetic systems (for which an inward proton flux is always observed) only proton binding reactions should occur. In order to demonstrate experimentally a proton loop mechanism, however, proton-releasing reaction should be studied as well, which should take place in those systems within the inner lumen of the vesicles. This difficult problem has been attacked so far only in chloroplasts, firstly by JUNGE *et al.* [47] for the study of the H⁺-releasing reactions of water photoxidation.

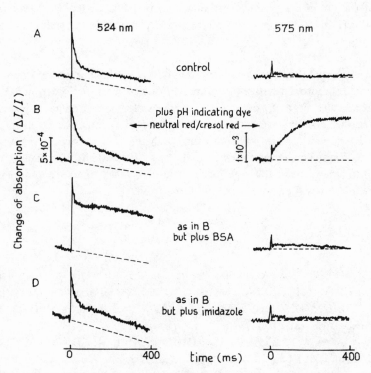

Fig. 13. — Single turnover flash response of neutral red or cresol red in a chloroplast suspension. Conditions: pH_{out} = 7.2, neutral red, 10 μM; cresol red, 30 μM; bovine serum albumine (C) 1.3 mg/cm³; imidazol (D), 1 mM. Average of 30 flashes fired at 0.1 Hz. (from Ref. 47).

For these studies the membrane permeating pH indicator neutral red has been employed; given its non-ionic structure this dye will distribute itself in both the inner and outer compartments, and within the lipid membrane core as well, and consequently its spectral signals will report on the pH variations in all these compartments. To obtain information specifically from the inner phase, the pH variations in the outer compartments were buffered out using high concentration of serum albumine in the suspending medium, which was acting as a non permeant outer buffer [47]. The resulting signals (Fig. 13) were therefore interpreted as due uniquely to pH variations in the inner lumen of the thylakoid, although, in this author's opinion, the possibility that the indicator dissolved in the membrane lipids could also contribute, cannot be ignored; in support to the interpretation offered originally, a strong inhibition of the signal by imidazol, a permeating buffer, was demonstrated. The results of these experiments demonstrated a good kinetic matching between the water oxidation steps and the H^+-releasing events within the inner lumen of the chloroplasts, and were therefore taken as evidence of good equilibration of the H^+ in the inner water phase and the intravesicular redox site for water oxidation [48].

Hardly any other example along this line is available in other systems, either respiratory or photosynthetic; this is mainly due to intrinsic experimental difficulties which are very hard to overcome and have so far prevented any advancement in this direction. It is however only by studying the fast kinetics of protolytic reactions, and by demonstrating its matching with the redox reaction velocity and its consistency with the thermodynamic properties of the electron transport catalysts, measured independently, that the mechanism of proton movement in energy transducing membrane will be elucidated.

References

[1]   H.W. HELDT, Ann. Rev. Plant Physiol. 32, 139 (1981).

[2]   M. KOIKE, L.J. REED and W.R. CARROLL, J. Biol. Chem. 238, 30 (1963).

[3]   H. FERNÁNDEZ-MORÁN, L.J. REED, M. KOIKE and C.R. WILLMS, Science 145, 930 (1964).

[4]   H. NAWA, W.T. BRADY, M. KOIKE and L.J. REED, J. Am. Chem. Soc. 82, 896 (1960).

[5]   L.J. REED and R.M. OLIVER, Brookhaven Symp. Biol. 21, 397 (1968).

[6]   L. SPATZ and P. STRITTMATTER, J. Biol. Chem. 248, 793 (1973).

[7]   M.J. ROGERS and P. STRITTMATTER, J. Biol. Chem. 249, 895 (1975).

[8]   M.J. ROGERS and P. STRITTMATTER, J. Biol. Chem. 249, 5565 (1974).

[9]   L. SPATZ and P. STRITTMATTER, Proc. Natl. Acad. Sci. U.S.A. 68, 1042 (1971).

[10] P. MITCHELL, Chemiosmotic Coupling and Energy Transduction, Glynn. Res. Bodmin, Cornwall, England, (1968).

[11] H.A. LARDY and S.M. FERGUSON, Ann. Rev. Biochem. 38, 991 (1969).

[12] A. BACCARINI-MELANDRI, R. CASADIO and B.A. MELANDRI, Curr. Top. Bioenerg. 12, 197 (1981).

[13] G.S. WILSON, Meth. Enzymol. 54, 396 (1978).

[14] D. ZANNONI, A. BACCARINI-MELANDRI, B.A. MELANDRI, E.M. EVANS, R.C. PRINCE and A.R. CROFTS, *FEBS Lett.* **48**, 152 (1974).

[15] P.L. DUTTON, M. ERECINSKA, N. SATO, Y. MUKAI, M. PRING, and D.F. WILSON, *Biochim. Biophys. Acta* **267**, 15 (1972).

[16] K.I. TAKAMIYA and P.L. DUTTON, *Biochim. Biophys. Acta,* **546**, 1 (1979).

[17] R.C. PRINCE and P.L. DUTTON, *Biochim. Biophys. Acta* **462**, 731 (1977).

[18] A. BACCARINI-MELANDRI and B.A. MELANDRI, *FEBS Lett.* **80**, 459 (1977).

[19] P.C. HINCKE and P. MITCHELL, *J. Bioenerg.* **1**, 45 (1970).

[20] D. WALZ, *Biochim. Biophys. Acta* **505**, 279 (1979).

[21] R.S. CAPLAN and A. ESSIG, *Proc. Natl. Acad. Sci. U.S.A.* **64**, 211 (1969).

[22] S. SCHULDINER, H. ROTTENBERG and M. AVRON, *Eur. J. Biochem.* **25**, 64 (1972).

[23] H. ROTTENBERG, *Meth. Enzymol.* **55**, 547 (1979).

[24] P.A. MICHEL and W.N. KONINGS, *Eur. J. Biochem.* **85**, 147 (1978).

[25] H. ROTTENGERG and T. GRÜNVALD, *Eur. J. Biochem.* **25**, 71 (1972).

[26] R. CASADIO and B.A. MELANDRI, *J. Bioenerg. Biomem.* **9**, 17 (1977).

[27] D.B. KELL, S.J. FERGUSON and P. JOHN, *Biochim. Biophys. Acta* **502**, 111 (1978).

[28] R. CASADIO, G. VENTUROLI and B.A. MELANDRI, *Photobiochem. Photobiophys.* **2**, 245 (1981).

[29] C.A. WRAIGHT, R.J. COGDELL and B. CHANCE, in *The Photosynthetic Bacteria,* R.K. CLAYTON and W.R. SISTROM, (Editors) Plenum, New York 1978, p. 471.

[30] J.B. JACKSON and A.R. CROFTS, *FEBS Lett.* **4**, 185 (1969).

[31] W. JUNGE and H.T. WITT, *Z. Naturforsch. Teil B* **23**, 244 (1968).

[32] W. LIPTAY, *Angew. Chem. Int. Ed. Eng.* **8**, 177 (1969).

[33] K. MATSMURA and M. NISHIMURA, *Biochim. Biophys. Acta* **459**, 483 (1977).

[34] A. BACCARINI-MELANDRI, R. CASADIO and B.A. MELANDRI, *Eur. J. Biochem.* **78**, 389 (1977).

[35] V. PICK and M. AVRON, *Biochim. Biophys. Acta* **440**, 189 (1976).

[36] B. CHANCE, M. BALTSCHEFFSKY, J. VANDERKOOI and W. CHENG, in *Perspectives in Membrane Biology,* S. ESTRADA and C. GITLER (Editors) Academic Press, New York (1974), p. 329.

[37] B. RUMBERG and H. MÜHLE, *Bioelectrochem. Bioenerg.* **3**, 393 (1976).

[38] J.B. JACKSON and A.R. CROFTS, *Eur. J. Biochem.* **10**, 226 (1969).

[39] J.B. JACKSON and P.L. DUTTON, *Biochim. Biophys. Acta* **325**, 102 (1973).

[40] H.T. WITT, *Biochim. Biophys. Acta* **505**, 355 (1979).

[41] M. SCHÖNFELD, M. MONTAL and G. FEHER, *Proc. Natl. Acad. Sci. U.S.A.* **76**, 6351 (1979).

[42] K.M. PETTY and P.L. DUTTON, *Arch. Biochem. Biophys.* **172**, 346 (1976).

[43] R.C. PRINCE and P.L. DUTTON, *FEBS Lett.* **65**, 117 (1976).

[44] K.M. PETTY, J.B. JACKSON and P.L. DUTTON, *FEBS Lett.* **84**, 299 (1977).

[45] W. AUSLÄNDER and W. JUNGE, *Biochim. Biophys. Acta* **357**, 285 (1974).

[46] K.M. PETTY, J.B. JACKSON and P.L. DUTTON, *Biochim. Biophys. Acta* **546**, 17 (1979).

[47] W. AUSLÄNDER and W. JUNGE, *FEBS Lett.* **59**, 310 (1974).

[48] P.L. DUTTON, *Meth. Enzymol.* **54**, 410 (1978).

[49] J.C. REED and D.J. COX, in *The Enzymes,* P.D. BOYER (Editor), Academic Press, New York, (1973) Vol. **1**, p. 218.

KINETICS IN BIOELECTROCHEMISTRY
PHOTOREDOX REACTIONS AT ELECTRODES AND IN SOLUTION

HERMANN BERG

Academy of Sciences of the GDR, Central Institute of Microbiology and Experimental Therapy, Department of Biophysical Chemistry, Jena, G.D.R.

Contents

1. Introduction to polarographic measurements of rate phenomena

The main aim of this chapter is to demonstrate the influence of light on biochemical redox reactions near an electrode surface or in the adjacent solution, coupled with or by the electrode reactions itself (transfer reactions, Durchtritts-reaktionen) [1-4]. In this respect, the metal or the semiconductor electrode works as a tool in photobiology on the basis of an electron exchange:

$$\text{Ox} + n\,\text{e}^- \underset{k_{e^+}}{\overset{k_{e^-}}{\rightleftharpoons}} \text{Red} \tag{1}$$

with the heterogeneous rate constants k_e, which depend on the electrode potential difference $\Delta U = U - U^\circ$ (U° = equilibrium potential):

$$k_{e^-} = k_{e^-}^\circ \exp\left(-\frac{\alpha\,n\,\text{F}\,\Delta U}{R\,T}\right) \tag{2a}$$

and

$$k_{e^+} = k_{e^+}^\circ \exp\left(\frac{(1-\alpha)\,n\,\text{F}\,\Delta U}{R\,T}\right) \tag{2b}$$

In the reversible case the standard potential U° is practically equal to the polarografic half wave potential $U_{1/2}$, because [Ox] = [Red] and $k_{e^-} = k_{e^+}$

$$U^\circ = U_{1/2} = \frac{R\,T}{n\,\text{F}} \ln \frac{k_{e^-}^\circ}{k_{e^+}^\circ} \tag{3a}$$

and according to NERNST

$$U = U^\circ - \frac{0,059}{n} \log \frac{[\text{Red}]}{[\text{Ox}]} = U_{1/2} - \frac{0,059}{n} \log \frac{I}{I_d - I} \tag{3b}$$

with I_d = diffusion current.

For an irreversible system ($k_{e^-} < 10^{-5}$ cm s^{-1}) the drop time, τ, the diffusion coefficient, D, and the transfer coefficient, k, must be taken into account:

$$U_{1/2} = \frac{2.3\,R\,T}{n\,F} \log \left[0.886\,k_{e^-}^{\circ} \left(\frac{\tau}{D} \right)^{1/2} \right] \tag{4}$$

Combination of (3) and (4) leads to the well known equation for the whole irreversible step:

$$\frac{I_d - I}{I} = \frac{k_{e^+}}{k_{e^-}} + \frac{1}{0.886\,k_{e^-}^{\circ}} \left(\frac{\tau}{D} \right)^{1/2} \exp \left(\frac{\alpha\,n\,F\,(U - U^{\circ})}{R\,T} \right) \tag{5}$$

with k_e values between 3 for $Cd^{2+} \rightarrow Cd$ and 10^{-5} for $Cr^{2+} \rightarrow Cr^{3+}$ at the mercury electrode.

This *transfer reaktion* limited by diffusion is modified by three types of chemical reactions in the vicinity of the electrode:
the preceding reaction

$$Ox' \underset{k'}{\overset{k}{\rightleftarrows}} Ox + n\,e^- \underset{k_{e^+}}{\overset{k_{e^-}}{\rightleftarrows}} Red \tag{6}$$

the parallel reaction (pseudomonomolecular)

$$Ox + n\,e^- \underset{k_{e^+}}{\overset{k_{e^-}}{\rightleftarrows}} Red \\ \underset{k}{\underline{\hspace{3cm}}} \overset{+}{} Ox' \tag{7}$$

the post reaction,
a) dimerization

$$2\,Ox + 2\,e^- \underset{k_{e^+}}{\overset{k_{e^-}}{\rightleftarrows}} 2\,Red^{\cdot} \overset{k_2'}{\longrightarrow} (Red')_2 \tag{8}$$

b) inactivation

$$Ox + n\,e^- \underset{k_{e^+}}{\overset{k_{e^-}}{\rightleftharpoons}} Red \xrightarrow{k'} Red' \tag{9}$$

In every case the differential equation for diffusion and the growing drop is extended for the chemical reactions, respectively:

in case (6) we have

$$\frac{d[Ox]}{dt} = D\,\frac{d^2[Ox]}{dx^2} + \frac{2\,x\,d[Ox]}{3\,t\,dx} + k\,([Ox'] - K\,[Ox]) \tag{10}$$

with $K = k/k'$ and with the ratio of mean kinetic current, I_k, to the diffusion current, I_d,

$$\frac{I_k}{I_d} = \frac{0.886\,(k\,\tau/K)^{\frac{1}{2}}}{1 + 0.886\,(k\,\tau/K)^{\frac{1}{2}}} \tag{11}$$

In case (7) we obtain

$$\frac{d[Ox]}{dt} = D\,\frac{d^2[Ox]}{dx^2} + \frac{2\,x\,d[Ox]}{3\,t\,dx} + k\,[Red] \tag{12}$$

and the ratio of the kinetic current to the diffusion current

$$\frac{I_k}{I_d} = 0.81\,(k[Ox']\,\tau)^{\frac{1}{2}}, \text{ approximately.} \tag{13}$$

In case (8) the half wave potential changes in *d.c.* polarography is given by

$$U_{\frac{1}{2}} = U^\circ - 0.36\,\frac{R\,T}{n\,F} + \frac{R\,T}{3\,n\,F}\,\ln\,([Ox]\,k_2'\,\tau) \tag{14}$$

In case (9) at high frequencies of *a.c.* polarography k_2' can be calculated from

$$(k_2')^{\frac{1}{2}} = \frac{1}{1.387\,(\tau)^{\frac{1}{2}}}\left(\frac{4.78\,I_d}{I} - 1\right)^2 \tag{15}$$

where I_d stands for the calculated diffusion current without introducing k'. For each mechanism examples will be given.

1.1.   The preceding reaction (6)

In a series of cancerostatic quinone derivates several compounds were in equilibrium with a quinone structure C (Fig. 1), which exhibits a more negative

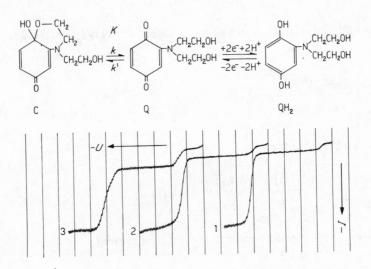

Fig. 1. — a) Preceding reactions and redox equilibrium of benzoquinone derivatives.
b) Polarogramms at pH 6 (1), 7 (2), 8 (3) first step: Q, second step: C.

irreversible reduction step [5]. Therefore the limiting current of the reversible positive quinone step is a kinetic one, which increases with temperature. For calculation of k the equation (11) is suitable if the equilibrium constant is determined by absorption spectroscopy. This was done as a function of temperature, because k was also determined by the relaxation time τ' in the ms range of the temperature-jump technique, in order to get thermodynamic data. We obtain

$$k = \frac{K}{\tau'(K + 1)} \quad \text{and} \quad k' = \frac{k}{K} \tag{16}$$

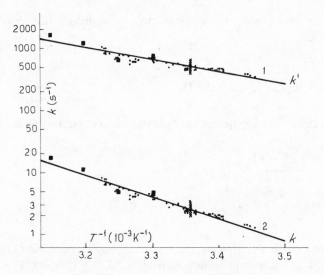

Fig. 2. — Arrhenius plots of rate constants (k and k') obtained from polarography and T-jump experiments at different concentrations. Polarography 1.1×10^{-4} M (■); 2×10^{-5} M to 2×10^{-4} M ($25\,^{\circ}$C) (x); T-jump 1 and 2×10^{-3} M (●).

The ARRHENIUS plot in Fig. 2 shows a rather good agreement between polarographic results from equation (11) and relaxation data from equation (16) in the low concentration range (10^{-4} M). The reaction enthalpy of $\Delta H = 29$ kcal mol^{-1} characterizes the ring opening to the Q-form as an endothermic process connected with an increase of entropy.

This was the first comparison between a polarographic and transient (T-jump) method.

1.2. The parallel reaction (7)

The application of equation (13) has been tested with cancerostatic active and inactive quinone derivatives [6, 7] with regard to reduction kinetics in the whole solution [8, 9]. A linear free energy relationship of the kind [10]

$$\Delta \log k = B\, \Delta U_{\frac{1}{2}} \tag{17}$$

was found even under microheterogenic conditions, which means that the quinones are reduced by palladium-sol hydrogen or by the dehydrogenases of microorganisms.

In solution the decrease of the quinone step by hydrogenation can be followed polarographically:

$$Q + [2H^\bullet]_{Pd \text{ (or bacteria)}} \xrightarrow{k} QH_2 \tag{18}$$

and at the electrode the kinetic current I_k is measurable resulting from:

$$QH_2 \xrightarrow{-2e^- - 2H^+} Q \tag{19a}$$

$$\begin{array}{c} \uparrow \qquad\qquad\qquad\qquad + \\ \underline{\qquad\qquad\qquad\qquad} 2H^\bullet_{Pd} \\ k \end{array} \tag{19b}$$

by analogy to equation (7). In the diagram according to equation (17) the normal quinones show a linear dependence with $B = 2.5 \text{ V}^{-1}$ (Fig. 3). The equilibrium quinones of Fig. 1, however, show lower kinetic currents, because k', is in the order of 60 s^{-1}. The projection of their k values to the line in Fig. 3 follow apparent $U_{1/2}$ which are between their half wave potentials of the first and second steps. In the case of *E.coli* reduction in solution (18), a B value of 4 V^{-1} was found. The activity range of dehydrogenases can be ascertained in this way.

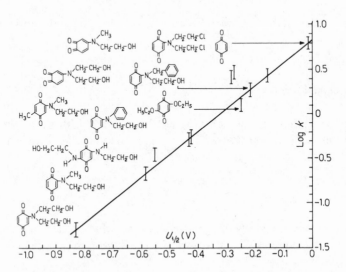

Fig. 3. — Dependence of $\log k$ from $U_{1/2}$ of a series of quinone derivatives.

1.3.   The postreactions

1.3.1. Dimerization (case 8).

— Dimerization by the taking up of one electron per monomer and one proton occurs very often, *i.e.* for pyridine nucleotides:

$$RN + e^- + H^+ \longrightarrow \overset{\bullet}{R}NH \tag{20a}$$

$$2\ \overset{\bullet}{R}NH \xrightarrow{\ k_2'\ } (RNH)_2 \tag{20b}$$

Table 1 shows results measured by cyclic voltammetry and peak height (CV-*I*), cyclic voltammetry and peak potential (CV-*U*), cyclic chronopotentiometry (CCP) and for comparison the higher value of pulse radiolysis and spectrophotometry (PR) as well as activation energies E_a and frequency factors A [4].

1.3.2. Inactivation (case 9).

— In the case of the reduction of a purine in aqueous solution the following steps were determined [16]. After the uptake of 2 e^- the carbanion becomes protonated (k') and the back reaction is blocked (Fig. 4). With aid

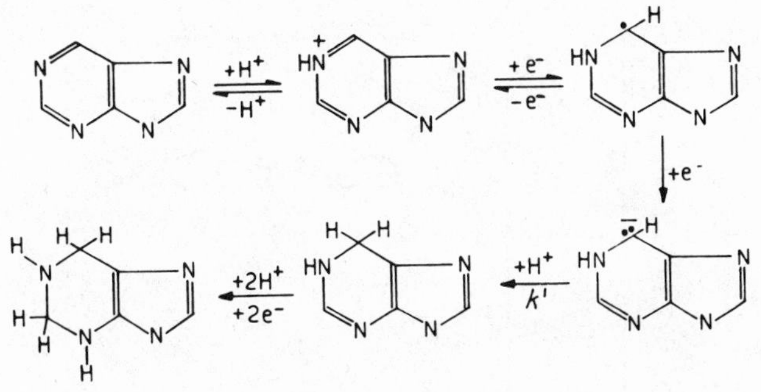

Fig. 4. — Mechanism of purine reduction according to P. ELVING.

of equation (15) k' can be determined as the pseudo first order rate constant at pH 3 to 5 (k_e^o_ exceeds 0.1 to 1 cm s^{-1}). In the case of pyrimidine this protonation k' is followed by slower deamination as shown in Table 2 [17].

Recently, we analyzed a special case of this scheme by a combination of coulometry and *d.c.* polarography of cancerostatic anthracycline antibiotics [18]. Following the large scale reduction by *d.c.* polarography a delay has been recorded in addition to the usual shift of the reversible reduction step (Fig. 5a). This delay corresponds to the consumption of 1 to 2 e^- before the hydroquinone is formed by

Table 1. — Dimerization rates of the free radicals derived from pyridine nucleotides

Compound	Solvent	Temp. °C	Method	k_2' M^{-1} s^{-1}	E_a kcal mol^{-1}	A M^{-1} s^{-1}	Ref.
NMN^+	H_2O, pH 5-9	30	CV-I	1.5×10^6	3.6	8.0×10^8	11
NAD^+	H_2O, pH 5	30	CV-I	2.2×10^6	9.0	7.0×10^{12}	12
	pH 7		CV-U	8.49×10^6			13
	pH 8		CCP	1×10^6			14
	pH 9		CV-I	2.4×10^6			12
			PR	5.6×10^7			15
$DNAD^+$	H_2O, pH 9	30	CV-I	1.7×10^6			12
$NADP^+$	H_2O, pH 5	30	CV-I	4.3×10^6	9.0	1×10^{13}	12
	pH 7		CV-U	5.45×10^{10}			13
	pH 9		CV-I	1.6×10^6			12

Table 2. — First order rate constants for pyrimidines and purines in aqueous medium

Compound	Carbanion protonation s^{-1}	Reduction product deamination
Pyrimidine (1,2-diazine)	5 x 10^4	
Cytosine (2-hydroxy-4-amino pyrimidine)	5 x 10^4	10 s^{-1}
CMP	5 x 10^4	3 s^{-1}
CpC	5 x 10^4	very slow
Purine (first 2 e⁻ addition)	4.9 x 10^3	
Purine (second 2 e⁻ reduction)	1.5 x 10^4	
Adenine (1-aminopurine)	2 x 10^5	
ApA	5 x 10^5	very slow

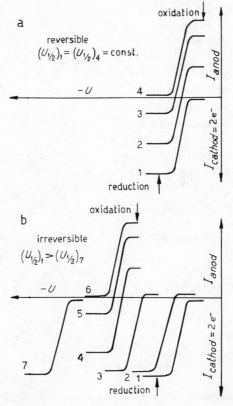

Fig. 5. — Large scale reduction schemes: a) normal quinone without chemical reaction; b) ECE mechanism resulting in sugar splitting.

2 e⁻ more (Fig. 5b). Fig. 6 shows the simplified mechanism of the first stages, namely the splitting of the sugar residues DA, RA or R in C-7-position. As a consequence the aglycone-quinone can be reduced reversibly at a slightly more negative potential. In organisms the sugar splitting is caused by dehydrogenases leading to a marked decrease of complex formation with nucleic acids and as a consequence to biological inactivity. For the new anticancer drug aclacinomycin A the electroactive dimer [compare (8)] has been identified by chromatography as the main product. Only iremycin with the sugar in C-10-position behaves as an usual anthraquinone (Fig. 5a), whereas β-rhodomycin exhibits a mixed behavior (Fig. 6). These polaro-

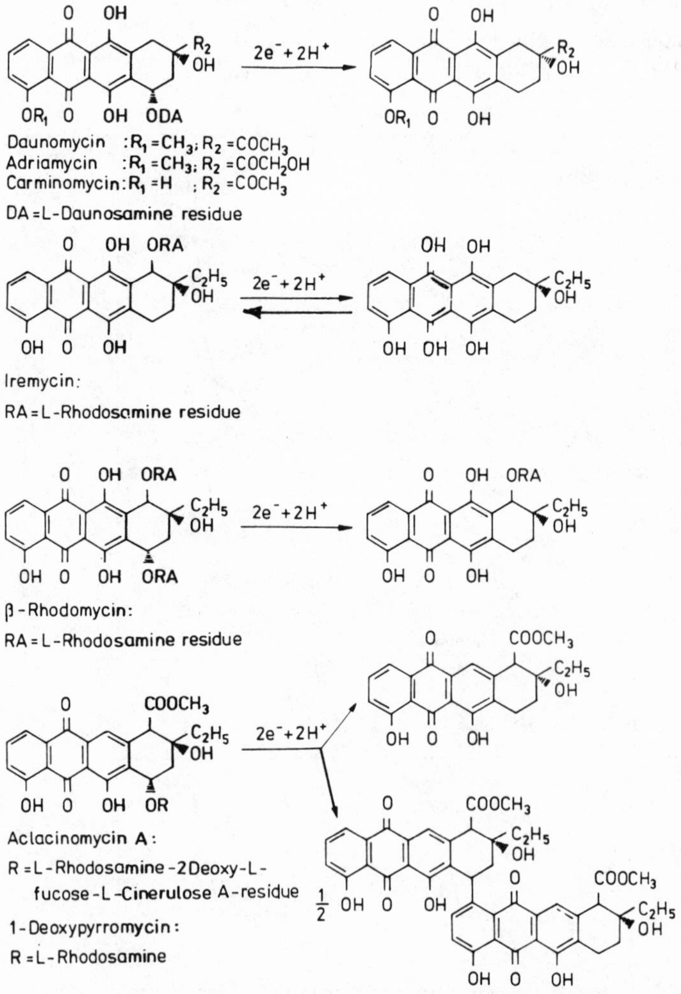

Fig. 6. — Four cases of quinone reduction (compare Fig. 5).

graphic results of ECE-type are in accordance with kinetic data of enzymatic inactivations.

Generally polarographic techniques are suitable for determining k_e values and chemical reactions coupled with electrode reactions up to k values of 10^9 M^{-1} s^{-1} (preceding protonation of weak acids) [19].

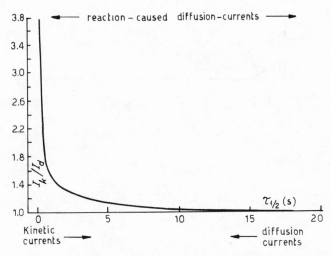

Fig. 7. — Dependence of ratio I_k/I_d from the half-life time $\tau_{1/2}$

However, there was a gap between the determination of k from I_k measurements and polarographic recording of slow reactions in solution as demonstrated in Fig. 7. Following the I-t-curves of single drops modified by reactions of half-life times between 1 and 20 s it is possible now [20] to measure such faster reactions by reaction-caused diffusion currents. Examples will be explained in Sections 2 and 3.

2.   Photoredox reactions of sensitizers at electrodes

From a principal point of view one has to distinguish between 6 cases, namely metal or semiconductor electrodes and absorption of light only by the electrodes or by the depolarizer or by both at high intensity [2, 21]. The main types of currents in photopolarography are drawn schematically in Fig. 8 [21-23]. In this Section metal and semiconductor conditions should be described in the presence of excited depolarizer (photodepolarizer sensitizer). The scavenging of photoelectrons will be presented in one example only.

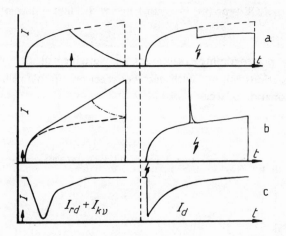

Fig. 8. — Scheme of photopolarographic current-time curves for permanent (left) and flash irradiation (right): a) photoreaction-caused diffusion currents (reduction); b) photokinetic current (oxidation); c) radical oxydation current as superposition.

2.1.   Photokinetic currents at metal electrodes (photopolarography)

Photokinetic currents can be produced by continuous or flash irradiation. In the first case stationary reactions are recorded as in the presence of an excess of Ox', H$^\bullet$, H$^+$ or other partners. In the second case dark reactions of photoproducts can be recorded, but not the transient concentration of singlet or triplet states (compare Section 2.2.).

The continuous irradiation has been performed with a low pressure mercury lamp surrounding a silica cell and the dropping mercury electrode (Fig. 9).

2.1.1. The preceding photoreaction. — By analogy to equations (6) and (10) the generation of Ox depends on the light intensity I_ν, the quantum yield ϕ_ν, and the extinction coefficient ϵ_{Ox}:

$$\frac{d[Ox]}{dt} = D\,\frac{d^2[Ox]}{dx^2} + \frac{2\,x\,d[Ox]}{3\,t\,dx} + \phi_\nu\,I_\nu\,\epsilon_{Ox}\,[Ox'] - k'\,[Ox] \qquad (21)$$

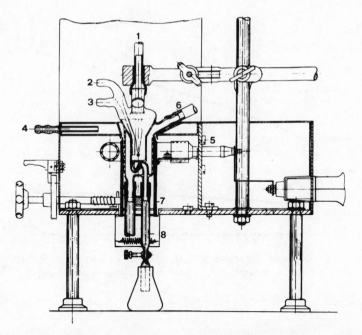

Fig. 9. – Silica vessel for polarographic measurements under permanent irradiation (U-burner) after start by the down-shift of a cylindrical shutter [10]. 1: d.m.e.; 2: gas inlet; 3: gas outlet; 4: coding device; 5: U-burner; 6: filter solution outlet; 7: cylindrical shutter; 8: solution outlet.

With $k_\nu = \phi_\nu\, I_\nu\, \epsilon_{Ox}$ for the relation of this photokinetic current $I_{k\nu}$, and neglecting k', one obtains

$$\frac{I_{k\nu}}{I_d} = \frac{0{,}886\,(k\,\tau)^{\frac{1}{2}}}{1 + 0{,}886\,(k\tau)^{\frac{1}{2}}}$$

2.1.2. The parallel photoreaction. – By analogy to equations (7), (11), and (19) respectively the following system has to be solved for quinones or reversible drugs (actinomycin for instance):

$$QH_2 - 2\,e^- - 2H^+ \longrightarrow Q$$
$$\uparrow \qquad k_\nu \qquad\qquad \downarrow h\nu \qquad\qquad (23)$$
$$\underline{\qquad\qquad} RH_2 + Q^*$$

$$\frac{d[Q]}{dt} = D\,\frac{d^2[Q]}{dx^2} + \frac{2\,x\,d[Q]}{3\,t\,dx} - \phi_\nu\, I_\nu\, \epsilon_Q\,[Q] \qquad\qquad (24)$$

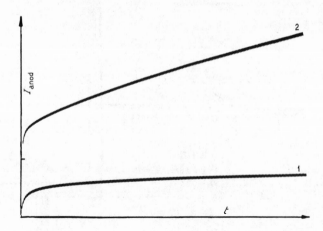

Fig. 10. — Diffusion current for (1) oxidation of anthraquinone-2-sulfonic acid and (2) photokinetic current (40 % *iso*propanol) by permanent irradiation.

With $k_v = \phi_v I_v \epsilon_Q$ one obtains

$$\frac{I_{kv}}{I_{d,\mathrm{QH_2}}} = 0.81 \, (k_v \, [\mathrm{RH_2}] \, \tau)^{\frac{1}{2}} \tag{25}$$

With Q = anthraquinone-2.6 disulfonic acid and $\mathrm{RH_2}$ = *iso*propanol (40 % in phosphate buffer) Fig. 10 shows I_d of $\mathrm{QH_2}$ in the dark and I_{kv} in the light. Since the equipment in Fig. 9 enables a fast start of the photoreduction:

$$\mathrm{Q} \xrightarrow{\ h\nu\ } \mathrm{Q^* + RH_2} \xrightarrow{\ k_v\ } \mathrm{QH_2} \tag{26}$$

in the whole solution a comparison between stationary results from I_{kv} with $k_v = 1.40 \ \mathrm{s^{-1}}$ is possible (see Section 3 and Fig. 7).

The measurement of the photokinetic current has the advantage of checking the photostability of reversible dyes and drugs, because the photodamage of reaction partners is easily recognized by the fading of I_{kv}.

Working with the flash irradiation in the μs range disturbances of I-t-curves are recorded only during the drop life (Fig. 11) [24]. This technique is successful for direct determination of dimerization of free radicals for instance of ketones (Fig. 11b) [25-27].

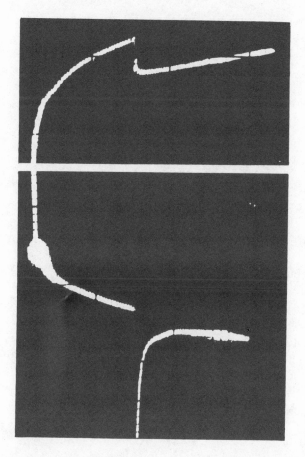

Fig. 11. — Flash irradiation within the drop-time of anthraquinone-2-sulfonic acid (top trace) and of the corresponding hydroquinone (bottom trace)

The use of stationary kinetic currents is very advantageous, however, it is only applicable for reversible or quasi-reversible electron transfers. In the irreversible case, on the other hand, the electron scavenger ability of substances in the vicinity of the electrode can be tested, if irradiation at shorter wavelength is performed (photoelectron emission).

Excited electrons in the electrode (*hot electrode*) [21]

$$(e^-)_{electrode} \xrightarrow{h\nu} (e^-)^*_{electrode} \rightsquigarrow$$

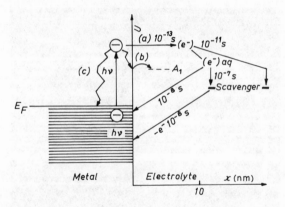

Fig. 12. — Scheme of electron emission from a metal (E_F-Fermi level).

can reduce an acceptor A_1 (event *b* Fig. 12) or can be emitted (event *a* Fig. 12 and Fig. 13). The following reactions are shown in Fig. 12. The measurable photoresidual current $I_\nu < I_e$ (emission current) in the absence of a scavengers (clean buffer solution) is rather low [21]. Effective electron scavengers are H_3O^+ [21, 31], N_2O [28, 30, 32], NO_2^- [28, 30], working as amplifiers for the measurement of photocurrents, because not all hydrated electrons $(e^-)_{aq}$ are able to return into the electrode:

$$\begin{array}{c} \rightsquigarrow e^- + H_3O^+ \searrow \\ \downarrow \qquad\qquad\qquad 2H^\bullet + 2\,H_2O \rightarrow H_2 \\ (e^-)_{aq} + H_3O^+ \quad k_2 = 2.4 \times 10^{10}\ M^{-1}\ s^{-1} \end{array} \qquad (27)$$

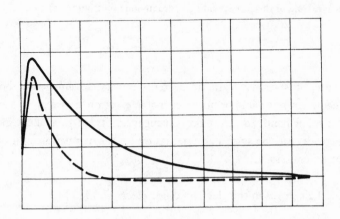

Fig. 13. — Electron emission after flash (- - -), division (corrected) 10 μs.

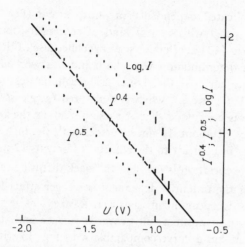

Fig. 14. — Experimental photo currents interpreted in terms of various theoretical laws: $I^{0.5}$ [31], $I^{0.4}$ [2], log I [21, 32], d.m.e. in 0.1 M KF solution.

In these *Photo-transfer-reaktions* the variables of equation (2) have to be substituted by $k_{e^-}^*$, $k_{e^+}^*$, k_e^{0*}, α^* [21] and the photocurrent can be linearized in the best way by the $I^{0.4}$ relationship [2] (Fig. 14).

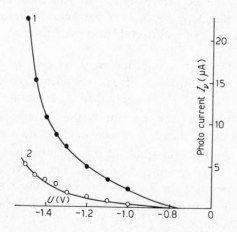

Fig. 15. — Photo current I_ν at a h.m.d.e. covered with DNA (1) and with the buffer (2) alone.

A more complicated dependence of photocurrent (Photo-transfer-reaktion) on U has been measured with a new kind of scavengers, namely adsorbed biopolymers. For proteins [33] as effective scavengers because of their sulfide, carbonyl and amino groups a determination with the h.m.d.e. is worked out for concentrations less than 0.5 μg/cm^3. At the d.m.e. using light pulses native DNA shows a maximum at -1.0 V (vs. s.c.e.), whereas denatured (single stranded) DNA has it at -1.5 V [34]. It was concluded that at these potentials the amount of orientated loops in the distance of about 10 nm coincides with the highest concentration of $(e^-)_{aq}$ for scavenging in contrast to the adsorbed segments (2 nm layer), where the stationary $(e^-)_{aq}$ concentration is low and the back diffusion is easier. This characteristic difference in the potential dependent scavenger effect between double and single stranded DNA is an argument against at least extensive surface denaturation of native DNA. In the case of a fully covered surface and stationary irradiation (360 nm) of the h.m.d.e. a curve comparable with Fig. 15 without maximum has been measured [35, 36]. In contrast to the d.m.e. [34] the distribution between once adsorbed segments and loops is not changed so drastically with potential [36].

2.2. Exchange currents of excited states at semiconductor electrodes

One of the important problems in photobiology is the determination of the electron exchange potentials of excited states for instance those of chlorophylls in the pathway of photosynthesis. According to the terms in metals Fig. 16(a) [37] in principle the measurement would be possible. That is expected because the excited level of Red* is higher than the Fermi-niveau E_F. Unfortunately by process (b) a competition of defect-electron (h^+)-injection into the free orbital of Red occurs and furthermore a quenching reaction (c) takes place. Therefore we were not able to measure $U_{1/2}^*$ of sensitizers [38] at the d.m.e.

The conditions for electron injection are suitable in the case of semiconductors. An energy transfer is forbidden, because the energy gap ΔE between valence band E_V and conduction band E_C is higher than the excitation energy $h\nu$ for Red (Fig. 17) [37]. The defect-electron injection is also excluded because the ground state energy of Red is higher than E_V. If such an electrode is polarized anodically, electrons move from Red* to the inside of the n-semiconductor (ZnO, CdS). The behavior of Ox and Ox* is opposite. At the p-semiconductor GaP as the cathode [39], electrons can move from the E_V level into the ground state (S_0) of Ox* at a more positive potential than for Ox, where electrons have to jump from the E_C to the S_1 orbital of the acceptor.

It is now possible to enhance the oxidation current of eosin, pseudoisocyanine, chlorophyll and other excited donors by donors in the groundstate as allythiourea,

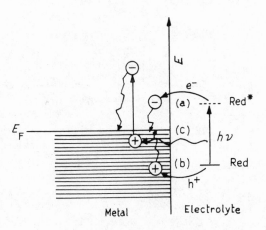

Fig. 16. — Electron exchange of excited state Red* at a metal electrode.

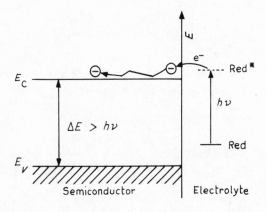

Fig. 17. — Electron exchange of excited state Red* at a semiconductor (E_C-valence band).

ascorbic acid, hydroquinone according to the following mechanism (28) called supersensitation:

$$D \xrightarrow{h\nu} D^* + QH_2 \xrightarrow{k_2} [D^*\text{-}QH_2] \longrightarrow e_c^- + D + QH^\bullet + H^+ \tag{28}$$

The electron that jumps into the conduction band is provided by the donor QH_2 in the charge transfer complex [40]. A second way is demonstrated in Fig. 18 [42] for chlorophyll and its reduction of the S_0 state:

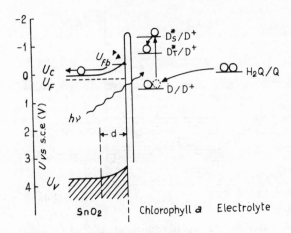

Fig. 18. — Supersensitation of chlorophyll a at an SnO_2-electrode (U_{fb} = flat band) by QH_2.

$$D^+ + QH_2 \xrightarrow{k_2} D + QH^\bullet + H^+ \tag{29}$$

A decrease of photocurrent on the other hand was found [42], if Q is present in solution (Fig. 19), because of the quenching reaction

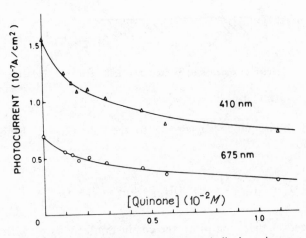

Fig. 19. — Inhibition of electron injection of chlorophyll a by quinone at two wave lengths.

$$D^* + Q \xrightarrow{k} D^+ + Q^{\cdot -}$$

The oxidation potentials of D_T and D_S of chlorophyll-a (Fig. 18) can be obtained by photocurrent-potential curves at semiconductors of different conduction band potentials in comparison to triplet (T) and singlet (S) excitation energies (1.26 eV and 18.6 eV). The difference between ground state oxidation potential $U_{\frac{1}{2}} = + 0.59$ V (s.c.e.), $U_{(T)}\frac{1}{2} = -0.67$ V and $U_{(S)}\frac{1}{2} = -1.27$ is higher than 1 V, which is the case also for reduction potentials [47, 58]. Types of linear free energy

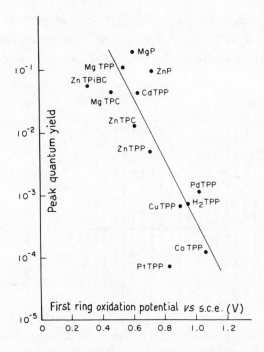

Fig. 20. — Linear relation for metalloporphyrins [41].

relationships for p-like porphyrins adsorbed on Al (Al_2O_3) were found for the quantum yield of electron production and the redox potential [41] (Fig. 20), and for the competitive oxidation parallel to water at TiO_2 of a series of reducing agents [42] (compare Section 4).

3. Photoredox reactions of sensitizers in solution

Slow photoreductions or oxidations [equation (26)] can be recorded by fixing a constant potential in the diffusion current region (Fig. 21) [10]. The same is possible for following dimerization reactions [10, 27] of free redicals after the flash (Fig. 22). For shorter half-life time of free radicals during continuous irradiation the diffusion-current caused by the photoreaction at the d.m.e., I_{pd}, can be measured [20] by:

$$I_{pd} = I_d [\exp(-k_\nu t)] (t - t_0 - u\,\tau)^{\bar{\alpha}} \tag{30}$$

with $k_\nu = \phi_\nu I_\nu \epsilon_{rad}$ and t = irradiation time, t_0 = time between start of the drop and start of irradiation, ϕ_ν = quantum yield, u = 0, 1, 2, 3 . . . for the first, second drop, ϵ_{rad} = extinction coefficient of the radical and $\bar{\alpha}$ = mean exponent of I_d-t curve.

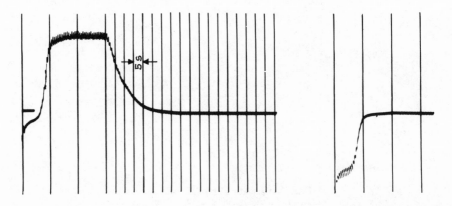

Fig. 21. — Permanent irradiation of anthraquinone-2-sulfonic acid in solution (40% *iso*propanol). Left: reduction current, right: oxidation current, time division 5 s.

For a series of photoreductions of benzoquinones according to equation (26) an adequate relation to equation (17) was found [21]:

$$\Delta \log \phi_\nu = B^* \Delta U_{\frac{1}{2}} \tag{17a}$$

with $B^* = 2.2 \text{ V}^{-1}$. Hence, it can be concluded that the half wave potentials of excited states $U_{\frac{1}{2}}^*$ should have differences that are similar to those of the ground states. For triplet quenching rate constants of acridine orange modified by a series

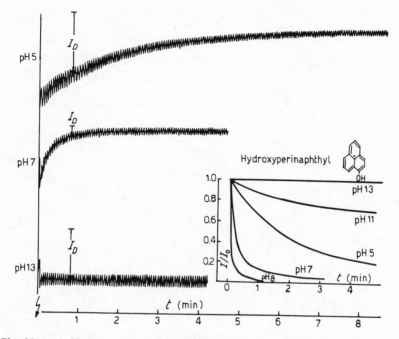

Fig. 22. — Oxidation currents of hydroxyperinaphthyl free radical after flash at pH 5-7-13, 40 % *iso*propanol. Insert: graphs relative mean current *vs.* time curves at different pH.

of electron donors a similar relation was found [43]. This result will be stressed for the photodynamic action [44, 46-48] on nucleic acids and their bases. After excitation of the sensitizer two main mechanisms I and II are possible as shown for guanidine in Fig. 23.

3.1. The photodynamic radical mechanism (Type I)

The sensitizer (methylene blue, acridine orange, thiopyronine) forms external and internal complexes with nucleic acids (Fig. 24), depending on the ratio of phosphate concentration to dye concentration. As shown by the spectra in Fig. 25 the high values P/D are typical for the internal binding (intercalation) of thiopyronine. Intercalation decreases the rate of side reactions of the transients $D_{(T)}$ (triplet) and $D^{\cdot+}$ (positive radical) (Fig. 26) and hence increases the probability to oxidize guanine (Fig. 23) to CO_2, parabanic acid and guanidine, which are the same endproducts as for the electrochemical oxydation [49].

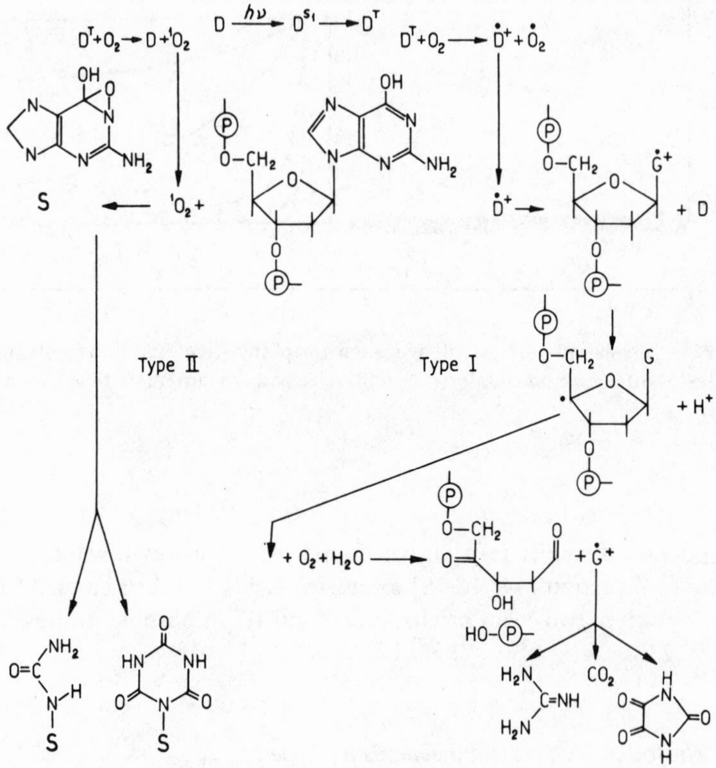

Fig. 23. — Scheme of two types of photodynamic oxidation of guanosin.

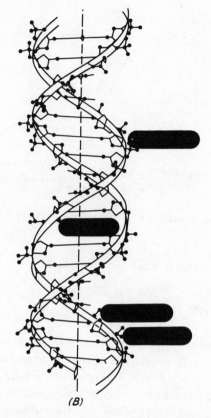

Fig. 24. — External and internal binding of a ligand to DNA (B-form).

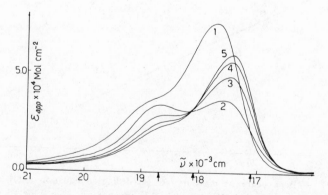

Fig. 25. — Absorption spectra of thiopyronin and titration with DNA, for different P/D values. 1) 0.0; 2) 8.02; 3) 16.03; 4) 24.05; 5) 32.07.

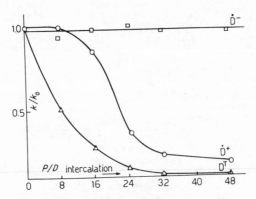

Fig. 26. — Relative reaction rates of transients of thiopyronin as a function of P/D.

The most important intermediate reaction is

$$D^{\cdot +} + G \xrightarrow{\;\bar{k}\;} G^{\cdot +} + D \tag{31}$$

where $\bar{k}$ depends on the difference in the redox potentials of $D/D^{\cdot +}$ and $G/G^{\cdot +}$. As shown in Fig. 27 with some bases a relation corresponding to equation (17) has a rather low value of $B = 1.06\ V^{-1}$. For practical purposes one can estimate the photodynamic effectiveness of sensitizers against donors by determining the difference in the redox potentials provided that the radical mechanism dominates [58].

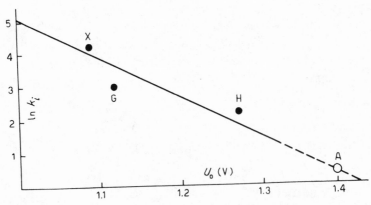

Fig. 27. — Dependence of reaction rate [according to equation (31)] between thiopyronin radical and X-xanthene, H-hypoxanthene, G-guanine, A-adenine (U_0 normal oxidation potential).

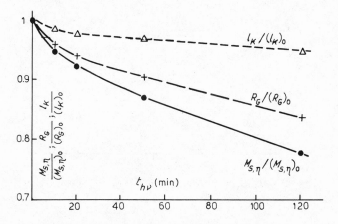

Fig. 28. — Change of molecular weight (M), radius of gyration (R_G) and KUHN length (l_K) during photodynamic degradation of DNA.

In the course of guanine degradation the sugar-phosphate backbone is broken, (Fig. 23), and it has to be repaired in the living cell [54]. Not only do single strand breaks occur, but also double strand breaks do as well, diminishing the molecular weight and radius of gyration from the beginning of irradiation (Fig. 28). The decrease of segment length according to KUHN indicates an increase of flexibility of the DNA chain. The reason why a double strand break is possible may be a successive electron-exchange (31) within an intercalation place between GC base pairs [45, 50, 56, 57]. Besides the nucleic acid degradation a lysis of microbiological cells is possible by photodynamic cell envelope damage [51, 60].

3.2. The photodynamic singlet oxygen mechanism (Type II)

While the transients of type I (Section 3.1.) can be recorded by fast spectroscopy, the direct determination of singlet oxygen $^1O_2$ is very difficult. Indirect analysis by blocking the reaction with NaN_3 is rather uncertain because the singlet state of the sensitizer responsible for $D^{\bullet+}$ generation is also quenched (Fig. 29).

In most cases sensitizers such as 6-mercaptopurine, toluidine blue, eosin, proflavine, hematoporphyrine do not form strong complexes with the substrates and they generate $^1O_2$ by energy transfer from the singlet state to oxygen [44, 48]. $^1O_2$ can freely diffuse and react with a lot of substrates at distant sites leading frequently to measurable amounts of peroxides.

For the photodynamic action of anthraquinones AQ a somewhat modified mechanism seems responsible [52, 53, 55]:

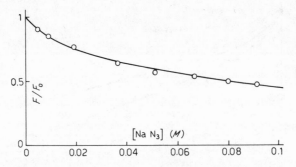

Fig. 29. — Quenching of fluorescence of thiopyronin by NaN_3.

$$AQ \xrightarrow{h\nu} AQ^* \tag{32a}$$

$$AQ^* + H_2O \longrightarrow AQH^\bullet + OH^\bullet \tag{32b}$$

$$2\,AQH^\bullet \longrightarrow AQ + AQH_2 \tag{32c}$$

$$AQH_2 + O_2 \longrightarrow AQ + H_2O_2 \tag{32d}$$

and $OH^\bullet$ or H_2O_2 cause strand breaks and inactivation of bacteria and cancer cells.

Hydrogen peroxide can be generated also by irradiation of powdered cadmium sulfide in aqueous air-saturated solutions, and thiols are oxidized to diols (Table 3) [59]. In the future, semiconductors may play a role in heterogeneous photodynamic effects (compare Section 4.1.) (Table 3).

4.   Photoredox reactions in microbiological and medical fields

Modelling and applications of photoredox reactions are also closely related to bioelectrochemistry.

4.1.   Models for biological photosystems

4.1.1. A chlorophyll-metal system (Fig. 30). — This arrangement [63] is like that for supersensitization (Section 2.2.), except that a platinum surface is used instead of a semiconductor. The amplitude of photopotential is highest with the mediator methylviologen for electron transport from excited chlorophyll to Pt. The albumin layer, A, prevents excitation energy transfer on the one hand, but allows electron migration on the other.

Table 3. — The effects of superoxide dismutase (SOD) and of ethanol on the quantum yields (ϕ_{ν}) of oxygen uptake during the cadmium sulfide-sensitized photooxidation of various biological substrates. The ratios of hydrogen peroxide produced to oxygen consumed (H_2O_2/O_2) are also shown for some substrates

Substrate	ϕ_{ν}			H_2O_2/O_2
	None	SOD	2 M Ethanol	None
Cysteine	0.021	0.012	0.019	0.02
Histidine	0.0011	—	—	0.21
Methionine	0.00012	—	—	—
Tryptophan	0.00040	—	—	0.40
Tyrosine	0.00026	—	—	—
Glutathione (reduced)	0.0037	0.0019	0.0035	0.02
Penicillamine	0.014	0.0046	0.015	0.08
Cysteamine	0.024	0.012	0.026	0.04
Dithiothreitol	0.013	0.013	0.011	0.05
Epinephrine	0.0046	0.0017	0.0027	1.63
EDTA	0.0063	0.0030	0.0064	0.75

4.1.2. A regenerative cell for solar energy conversion (Fig. 31). — A combination of a semiconductor electrode and a metal electrode is suitable for a solar cell [39, 64]:

$$\text{anodic process} \quad Red + p^+ \longrightarrow Ox \tag{33}$$

$$\text{cathodic process} \quad Ox + e^- \longrightarrow Red \tag{34}$$

4.1.3. Photoelectric simulation of electron pumping process. — This simulation during photosynthesis by means of a chlorophyll-layer-semiconductor photoanode and a chlorophyll-layer metal photocathode is illustrated by Fig. 32 [65] with inverse rectifying property. The external circuit works as the *in vivo* electron transfer chain across the thylakoid membrane.

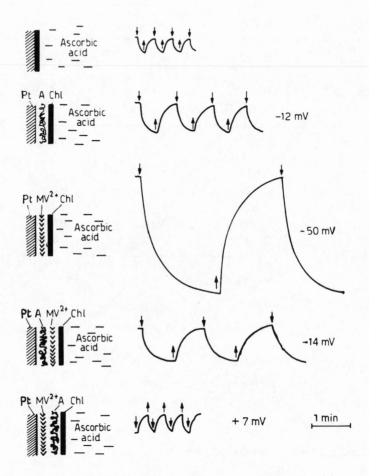

Fig. 30. — Combinations of layers and response of photopotential.

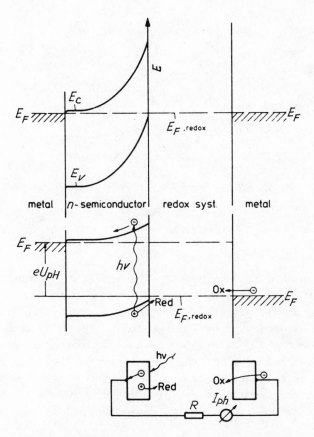

Fig. 31. — Element semiconductor-metal for sun energy conversion.

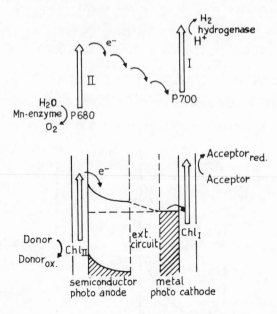

Fig. 32. — Electrochemical model for both chlorophyll systems.

4.1.4. Two compartment coupling water splitting (Fig. 33) by a membrane. — Dye D is a photochemical reducing agent which feeds electrons one at a time to the charge storage catalyst, M. When M has received two electrons, it is capable of catalyzing the reduction of two H^+ ions to H_2. The dye A is a photochemical oxidizing reagent which removes electrons one at a time from the charge-storage catalyst N. When N has lost four electrons, it is capable of catalyzing the oxidation of two water molecules to O_2. The membrane permits D^+ and A^- to react to reform D and A and allows protons to migrate from left to right to balance the charge [61].

4.1.5. Chlorophyll inside of liposomes. — Under unaerobic conditions the photo-reduction of $[Fe(CN)_6]^{3-}$ in the outside solution was enhanced by the presence of proton carrier such as carbonylcyanide m-chlorophenylhydrazone (Fig. 34) [62]:

$$Chl^* + [Fe(CN)_6]^{3-} \longrightarrow Chl^{\cdot+} + [Fe(CN)_6]^{4-} \tag{35a}$$

4.1.6. The central role of the electric field in photosynthesis. — If was analysed by H. WITT and his school [66]. The corresponding effect of light and external field strength on the yield of ATP formation is surprising.

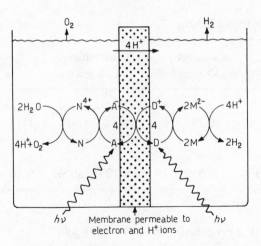

Fig. 33. — Electrochemical compartment model.

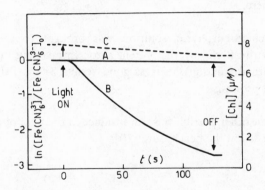

Fig. 34. — Liposome model for chlorophyll action: A without proton carrier; B with proton carrier; C bleaching of Chl.

4.2.   Photodynamic treatment against cancer and psoriasis

The photodynamic action according to type I against nucleic acids *in vitro* [50, 52] was applied to skin tumors of mouse, which had been induced by intra-dermal injection of 0.1 mm³ ascites (10^6 cells) dispersion [67]. The sensitizer (methylene blue, thiopyronine) in dimethylsulfoxide-water (1:1) was injected in the tumor region, which was then irradiated by a 750 W tungsten lamp 4 times for 45 min each day. The best results in healing were found with thiopyronine. They are collected in Table 4.

Table 4. — Photodynamic healing

Irradiation dates	Sensitizer mg/0.2mm^3	Cured mice Irrad. mice	Control	Irrad. effect
5-8 day (expansion stage)	0.12	9/12 = 75%	0/12 = 0 %	75 %
	0.25	12/12 = 100%	1/12 = 8.3 %	92.7 %
8-11 day (prenecrotic stage)	0.5	8/12 = 50 %	0/16 = 0 %	50 %

The type II mechanism, especially with soluble hematoporphyrin is also highly effective and has been applied later on to cancer patients [68] successfully. With the aid of light tranducers and laser-light this photodynamic treatment has been extended inside the body, too. The pioneering work of TH. DOUGHERTY of the ROSWELL Park Memorial Institute is most important for the introduction of this method elsewhere.

So some hematoporphyrin solution was brought to the Dermatological Hospital of the Charité Berlin by the present autor. Intravenous injection (1 mg/kg) and long wavelength irradiation yields as good results as the PUVA-therapy with 8-MOP [69] against psoriasis.

There are some other possibilities, for instance in the field of antivirus therapy, which will be discussed in the near future [70].

References

[1]   J. HEYROVSKY and J. KUTA, *Grundlagen der Polarographie*, Akademie Verlag, Berlin (1965).

[2]   Yu. GUREVICH, Yu. PLESKOV and Z.A. ROTENBERG, *Photoelectrochemistry*, Consultants Bureau, New York (1980).

[3]   H. GERISCHER, D. KOLB and J. SASS, *Adv. Phys.* **27**, 437 (1978).

[4]   P. ELVING and B. GRAVES, *Topics in Bioelectrochemistry and Bioenergetics*, G. MILAZZO (Editor), J. Wiley, Chichester, New York (1980), Vol. **3**, p. 1.

[5]   H. WEBER, W. FÖRSTER and H. BERG, *J. Electroanal. Chem.* **100**, 135 (1979).

[6]   H. WAGNER and H. BERG, *J. Electroanal. Chem.* **1**, 61 (1959).

[7]   H. BERG, E. BAUER and D. TRESSELT, *Adv. Polarogr.* **1**, 382 (1960).

[8]   H. KAPULLA and H. BERG, *J. Electroanal. Chem.* **1**, 108 (1959).

[9]   H. BERG and K. KRAMARCZYK, *Talanta* **12**, 1127 (1965).

[10] H. BERG, *Z. Chem. (Leipzig)* **2**, 237 (1962).

[11] C. SCHMAKEL, K. SANTHANAM and P. ELVING, *J. Electrochem. Soc.* **121**, 1033 (1974).

[12] C. SCHMAKEL, K. SANTHANAM and P. ELVING, *J. Am. Chem. Soc.* **97**, 5083 (1975).

[13] A. CUNNINGHAM and A. UNDERWOOD, *Biochemistry* **6**, 266 (1967).

[14] A. WILSON and D. EPPLE, *Biochemistry* **5**, 3170 (1966).

[15] E. LAND and A. SWALLOW, *Biochim. Biophys. Acta* **162**, 327 (1968).

[16] P. ELVING, S. PACE and J. REILLY, *J. Am. Chem. Soc.* **95**, 647 (1973).

[17] P. ELVING, *Topics in Bioelectrochemistry and Bioenergetics*, G. MILAZZO (Editor), J. Wiley, London (1976), Vol. **1**, p. 179.

[18] H. BERG, G. HORN and W. IHN, *J. Antibiotics (Japan)* **1981** in print.

[19] H.W. NÜRNBERG, *Fortschr. Chem. Forschg.* **8**, 241 (1967).

[20] H. BERG and H. KAPULLA, *Z. Elektrochem. Ber. Bunsenges. Phys. Chem.* **64**, 44 (1960).

[21] H. BERG, H. SCHWEISS, E. STUTTER and K. WELLER, *J. Electroanal. Chem.* **15**, 415 (1967).

[22] H. BERG and H. SCHWEISS, *Electrochim. Acta* **9**, 425 (1964).

[23] H. BERG, *Rev. Polarogr. (Kyoto)* **11**, 29 (1963).

[24] H. BERG, H. SCHWEISS and D. TRESSELT, *Exp. Tech. Phys.* **12**, 116 (1964).

[25] H. BERG and H. SCHWEISS, *Nature (London)* **191**, 1270 (1961).

[26] S. PERONE and J. BIRK, *Anal. Chem.* **38**, 1589 (1966).

[27] E. STUTTER, *J. Electroanal. Chem.* **50**, 315 (1974).

[28] G. BARKER, A. GARDNER and D. SAMMON, *J. Electrochem. Soc.* **113**, 1182 (1966).

[29] R. BALDWIN and S. PERONE, *J. Electrochem. Soc.* **123**, 1647 (1976).

[30] S. BABENKO, V. BENDERSKI, Y. ZOLOTOVITSKI and A. KRIVENKO, *J. Electroanal. Chem.* **76**, 347 (1977).

[31] P. DELAHAY and V. SRINIVASAN, *J. Phys. Chem.* **70**, 420 (1966).

[32] M. HEYROVSKY, *Nature (London)* **209**, 708 (1966).

[33] K. YAMASHITA and H. IMAI, *Bioelectrochem. Bioenerg.* **5**, 650 (1978).

[34] G. BARKER, *Lecture on Heyrovsky Mem. Congr. Polarogr. Prague* (1980).

[35] H. BERG, *Lecture on Heyrovsky Mem. Congr. Polarogr. Prague* (1980).

[36] H. BERG, *Der Formenwandel einer DNA-Doppelhelix in Lösung und seine molekularbiologische Bedeutung*, Akademie Verlag, Berlin (1981).

[37] H. GERISCHER, *Ber. Bunsenges. phys. Chem.* **77**, 771 (1973).

[38] H. BERG and F.A. GOLLMICK, *Coll. Czech. Chem. Commun.* **30**, 4192 (1965).

[39] R. MEMMING and H. TRIBUTSCH, *J. Phys. Chem.* **75**, 562 (1972).

[40] H. TRIBUTSCH and H. GERISCHER, *Ber. Bunsenges. phys. Chem.* **73**, 251, 850 (1969).

[41] F. KAMPAS, K. YAMASHITA and J. FAJER, *Nature* (*London*) **284**, 40 (1980).

[42] A. FUJISHIMA, T. INONE and K. HONDA, *J. Am. Chem. Soc.* **101**, 5582 (1979).

[43] E. VOGELMANN, W. RAUSCHER, R. TRABER and H. KRAMER, *Z. physik. Chem N.F.* (*Frankfurt*) **124**, 13 (1981).

[44] J. SPIKES, *Ann. Rev. Phys. Chem.* **18**, 409 (1967).

[45] H. BERG, F.A. GOLLMICK, H.-E. JACOB and H. TRIEBEL, *Photochem. Photobiol.* **16**, 125 (1972).

[46] H. SENGER (Editor), *The Blue Light Syndrome* Springer Verlag Berlin, (1980).

[47] L. KITTLER and G. LÖBER, *Photochem. Photobiol. Revs.*, K. SMITH (Editor), Plenum Press New York (1977).

[48] J. AMAGASA, *Photochem. Photobiol.* **33**, 947 (1981).

[49] H. BERG, *J. Electroanal. Chem.* **65**, 129 (1975).

[50] H. TRIEBEL, H. BÄR, H.-E. JACOB, E. SARFERT and H. BERG, *Photochem. Photobiol.* **28**, 331 (1978).

[51] H.-E. JACOB and M. HAMANN, *Photochem. Photobiol.* **28**, 331 (1978).

[52] H. BERG and H.-E. JACOB, *Z. Naturforsch.* **17b**, 306 (1962).

[53] H. BERG and H.-E. JACOB, *Photochem. Photobiol.* **4**, 55 (1965).

[54] H.-E. JACOB, *Photochem. Photobiol.* **14**, 743 (1971).

[55] H. BERG, *Rev. Polarogr.* (*Kyoto*) **14**, 351 (1967).

[56] H. BERG, F.A. GOLLMICK, H. TRIEBEL, E. BAUER, G. HORN, J. FLEMMING and L. KITTLER, *Bioelectrochem. Bioenerg.* **5**, 335 (1978).

[57] H. BERG, *Bioelectrochem. Bioenerg.* **5**, 356 (1978).

[58] L. KITTLER, G. LÖBER, F.A. GOLLMICK and H. BERG, *Bioelectrochem. Bioenerg.* **7**, 503 (1980).

[59] J. SPIKES, *Photochem. Photobiol.* **34**, 549 (1981).

[60] T. ITO, *Photochem. Photobiol.* **33**, 117 (1981).

[61] M. CHAN and J. BOLTON, *Photochem. Photobiol.* **34**, 537 (1981).

[62] K. KURIHARA, Y. TOYOSHIMA and M. SUKIGARA, *Biochem. Biophys. Res. Commun.* **88**, 320 (1979).

[63] B. KISELEV and Yu. KOZLOV, *Bioelectrochem. Bioenerg.* **7**, (1980) 247.

[64] T. MIYASAKA, T. WATANABE, A. FUJISHIMA and K. HONDA, *J. Am. Chem. Soc.* **100**, 6657 (1978).

[65] T. MIYASAKA and K. HONDA, *Photoeffects at Semiconductor Electrolyte Interface, Am. Chem. Soc. Symp.* (1980) p. 146.

[66] E. SCHLODDER and H.T. WITT, *Biochim. Biophys. Acta* **635**, 571 (1981).

[67] W. JUNGSTAND and H. BERG, *Stud. Biophys.* **3**, 225 (1967).

[68] Th. DOUGHERTY, J. KAUFMANN, A. GOLDFARB, K. WEISHAUPT, D. BOYLE and A. MITTLEMAN, *Cancer Res.* **38**, 2628 (1978).

[69] W. DIETZEL, H. MEFFERT, N. SÖNNICHSEN, *Dermatol. Monatsschr.* **167**, 617 (1981).

[70] *IX. Jena Symp.: Photodynamic and Electric Effects on Biopolymers and Membranes*, Sept. 20-25 (1981) Weimar.

APPLICATIONS OF ADVANCED VOLTAMMETRIC METHODS IN BIOELECTROCHEMISTRY

HANS WOLFANG NÜRNBERG

Institute of Applied Physical Chemistry
Chemistry Department, Nuclear Research Center
Juelich, Federal Republic of Germany

Contents

1.   Introduction

Certain modes of advanced voltammetry provide a significant methodological approach for a variety of research tasks and problems in bioelectrochemistry. The applications of voltammetry include the analysis and physicochemical characterization of biologically significant organic substances and biopolymers, studies on the fate and metabolism of drugs, on the interactions and effects that enzymes, mutagenic chemicals, ionizing radiation and toxic metals exert on biopolymers, the development of simple and reliable *in vitro* test procedures to monitor the resulting damages in nucleic acids, the investigation and mechanistic elucidation of enzymatic redox reactions and of photoredoxreactions, the performance of model studies on the biophysicochemical behavior of biopolymers at electrically charged interfaces, etc. Most applications of voltammetry have been hitherto performed *in vitro*. There is, however, reported in the literature, also a number of *in vivo* measurements, *e.g.* in brain chemistry [68-70].

A comprehensive treatment of the subject would be beyond the scope of this chapter. Therefore the potentialities of voltammetry will be featured by a variety of applications selected mainly from the respective research programs in the author's institute.

2.   General methodological aspects

The name *voltammetry* stems from *voltamperometry*. Voltammetric measurements consist in the recording of *current-potential-relationships* corresponding to the electrode process in which the studied substance is reduced or oxidized according to the fundamental relation:

$$S + n\,e^- + m\,H^+ \underset{\text{Oxidation}}{\overset{\text{Reduction}}{\rightleftharpoons}} P \tag{1}$$

As usually if organic substances are involved the reduction requires besides the uptake of electrons also the transfer of protons to the respective sites of the molecule.

In voltammetry the adjusted entity is always the electrode *potential U* and the measured parameter the resulting *current I*. Frequently the recorded current I corresponds to an electrode process, *i.e.* to the transfer of electrons through the interface electrode/solution. Then one has a *faradaic current* I_F. In principle each substance undergoes around its respective *redoxpotential* U_0 an electrode process. The value of U_0 is a function of the electrode material and basic parameters of the solution, particularly the pH.

The dependence of U_0 on the electrode material excludes certain types of organic compounds from voltammetric accessibility, if they contain no electroactive substituents which are reducible or oxidizable within the available potential range of the used working electrode. The most suitable working electrode for reductions is with respect to its high overvoltage for hydrogen evolution the Hg-electrode. It provides an applicable potential range from about $+0.2$ to -2.0 (s.c.e.). For oxidations up to $+2.0$ V (s.c.e.) working electrodes from graphite, carbon paste, glassy carbon and gold and sometimes silver are common. Table 1 provides an overview over the substance types accessible to voltammetry.

Although each electroactive substance has a corresponding reduction and/or oxidation potential at a given working electrode material and in the respective pH-range, frequently the potential ranges of several electroactive substances present in the solution overlap. This difficulty can be overcome by coupling voltammetry with an efficient separation procedure to be applied prior to the voltammetric determination of the individual components. The most efficient and important separation approach is high performance liquid chromatography (HPLC), but also other chromatographic techniques and solvent extraction have been used. Particularly elucidating can be also for mechanistic and interfacial studies the combination with spectroscopic methods.

Besides the faradaic current I_F related to an electrode process of the studied substance there flows always a further current component I_c. This *charging, or capacitive, current* I_c is connected with the alteration of the charge of the double layer existing at the working electrode. The electrical analogon of the double layer is a condenser and thus the double layer acts as a capacitance. Many organic substances have sufficiently large hydrophobic moieties to exert surface activity and become consequently adsorbed at the interface electrode/solution within a certain potential range. The resulting alteration in the dielectric of the double layer capacitance will induce the alteration of its charge and a consequent flow of I_c. With suitable modes of voltammetry this *non-faradaic* signal can be used as well for investigations. At this stage it is important to understand that the current I, which is the recorded signal in voltammetry, is in principle always the sum of two components according to:

$$I = I_F + I_c \tag{2}$$

Depending on the aim and type of the investigation, one selects voltammetric modes which make one of the current components practically negligible. In the

Table 1. — Substance types accessible to voltammetry (see also Ref. 5, 8, 9, 12)

a) *Reduction at mercury electrodes*
  — Nitro-, Nitroso-, Azo-, Azoxy-functions, Hydroxylamines, Oximes, Amine oxides
  — Aldehydes and many ketones, particularly if the carbonyl function is conjugated with double bonds, and the common derivated, *e.g.* Oximes, Hydrazones, Imines, Semicarbazones, Azomethines.
  — Peroxides
  — Halogen functions
  — Conjugated C=C-bonds and higher annelled Aromatic systems
  — Disulphides, Sulphoxides, Thioethers, Thiobenzophenones, Thiobenzamides, aromatic Thiocyanates and, if conjugated, Sulphones, Sulfonic Acids, Sulfinic Acids.
  — Heavy Metals, Cd, Pb, Cu, Zn, Ni, Co, Se(IV), As(III), Sb(IV) etc., and Organometallic Compounds
  — Most O-, N-, S-Heterocycles including mixed heterocycles containing these heteroatoms
    Among substances of immediate *biological significance:*
    — certain Coenzymes
    — most Vitamines, Hormones, Steroids and Antibiotics
    — Adenine and Cytosine Nucleotides, DNS and RNA
    — Proteins via S-S-bonds or non-faradaic responses

b) *Oxidation*
  — at mercury electrodes: Thiols, Thio-acids Xanthates, Proteins *via* SH
  — at graphite, glassy carbon, carbon paste and gold electrodes: Amines; Phenols; most Heterocycles
  Among substances of immediate *biological significance:*
  — Vitamines, Hormones, Steroides, Porphyrines, Antibiotics
  — Nucleotides and Nucleic Acids
  — certain Coenzymes
  — Proteins *via* SH

much more frequent case that the signal due to the electrode process of the respective substance is used one operates with methods where

$$I \approx I_F \tag{3}$$

while if the non-faradaic current is the desired signal modes have to be applied which are insensitive to I_F, so that the recording conditions refer to

$$I \approx I_c \tag{4}$$

Voltammetry is essentially a trace method and is for organic substances performed with bulk concentrations between 10^{-3} and 10^{-7} M or sometimes even less. The studied substance must be dissolved in a polar solvent with an $\epsilon \geqslant 30$. The most common and biologically particularly relevant solvent is water.

Besides the studied trace substances the solution has to contain a *supporting electrolyte* and usually a *buffer* to adjust a defined pH. Common supporting electrolytes are NaCl, KCl, LiCl and other alkaline metal salts. The function of the supporting electrolyte is to provide a sufficient conductance of the solution. The supporting electrolyte is present in the solution in an order of magnitudes higher concentration. As also the buffer components take over the function of the supporting electrolyte the added alkaline salt concentration is usually in the order of 0.1 to 0.5 M. Due to the substantial excess of the supporting electrolyte the transport of the studied trace substances to the working electrode occurs even if they are ions only by diffusion or convective diffusion. Moreover, the ohmic resistance of the solution and the concomitant IR-drop of the applied potential is kept small and tolerable, particularly if the todate common potentiostatic control of the working electrode potential is applied.

While the *general term* for the methods determining substances by the recording of their current-potential-relationship is *voltammetry* the term *polarography* is used if the working electrode is a *dropping mercury electrode* (d.m.e.). This name had been coined by the founder of this branch of electrochemistry, the Nobel Prize laureate J. HEYROVSKY, who published 1922 the first current-potential-curves recorded at the d.m.e. Meanwhile this approach has developed and diversified into a wide and still expanding spectrum of voltammetric methods many of them particularly adapted to certain applications in trace analysis of inorganic and organic substances or for the determination of a variety of their physicochemical and electrochemical parameters and properties.

The voltammetric approach has a number of *inherent basic advantages*.

It is a quasi-destructionless method, because the alteration of the bulk concentration of the studied substance remains virtually unchanged. The recording of the current-potential-relationship (voltammogram) consumes only 0.01 to 0.1 % of the bulk concentration, because only the substance amount in the immediate vicinity of the electrode is involved in the electrode process which has to occur to produce the voltammogram. Therefore, only a part of the substance amount in the diffusion layer of the working electrode is consumed and the effect on the bulk concentration remains undetectable even if the recording of the voltammogram is repeated many times in the same analyte.

A further practical advantage in comparison to other non-electrometric methods is that colour or turbidity of the solution cause no interference.

Voltammetry is fundamentally based on FARADAY's law according to which 1 Mol of substance undergoing an electrode process is equivalent to n x 96 500 C. The factor n is the number of electrons transferred in the elementary step of the electrode process of a molecule of substance (or for biopolymers a monomeric unit) through the interface electrode/solution. For organics n has usually values between 1 and 4. Sometimes it can be even larger. Due to the tremendous electrical charge equivalent to 1 mole of substance voltammetric methods combine high inherent accuracy, *i.e.* remarkably low tendency for systematic errors, with excellent determination sensitivity which is in many case superior to any or most other methodological alternatives. Precision is also good and amounts according to the level of the bulk concentration to relative standard deviations (RSD) between 1 and 20 % (at the determination limit in the ultra trace range).

Voltammetry is primarily a methodology focussed on the quantity of trace substances and not on their identification. Therefore, as already mentioned, mixtures of the organic trace substances to be determined require frequently the prior application of an efficient separation procedure, *e.g.* HPLC.

The current-potential-relationship resulting in voltammetry of the respective trace substance has usually the shape of a step or a peak (Fig. 1). The corresponding limiting current I_{lim} or peak current I_p is always proportional to the bulk concentration. There are cases, particularly in bioelectrochemical studies, where I_{lim} or I_p correspond immediately to the concentration adsorbed at the working electrode surface, but then this interfacial concentration is related to the bulk concentration by the adsorption isotherm.

The evaluated voltammetric response, I_{lim} or I_p, being proportional to the bulk concentration of the studied substance is frequently *diffusion controlled*.

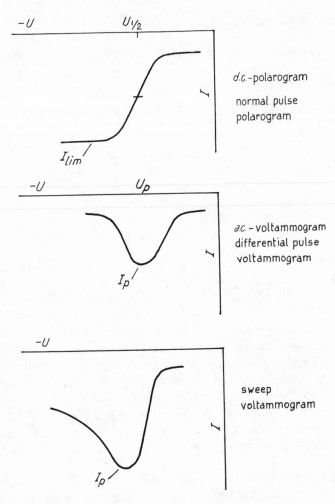

Fig. 1. — Types of voltammograms obtained with different voltammetric modes. I_{lim} limiting current, I_p peak current; $U_{1/2}$ half wave potential, U_p peak potential. I_{lim} and I_p are always proportional to bulk concentration

There are, however, cases where I_{lim} or I_p are partially also controlled by the *kinetics* of chemical reactions prior to the electrode process, *e.g.* proton transfer reactions, dehydration steps or various kinds of chemical transformations. This makes voltammetry a suitable method to study the kinetics of fast reactions in solution and also in the diffuse double layer part of charged interfaces, represented by the working electrode, under the influence of the electric field gradient, if the respective chemical reaction can be coupled to a suitable subsequent electrode reaction. An example is given by extended studies with a special voltammetric pulse method using μs-pulses on the dissociation and recombination kinetics of carboxylic acids a topic of general relevance to proton transfer processes in biological systems [1-3].

Of particular significance with respect to bioelectrochemical aspects is the conformational and structural behavior and fate of *biopolymers* and their monomeric units at charged interfaces, which occur in form of membranes and enveloping proteins for nucleic acids and nucleotides in the living cell. With respect to the fundamental physicochemical and electric parameters the interface electrode solution constitutes a valuable and easily manipulable model system.

For corresponding *interfacial studies* one has basically two alternatives. Either one makes use of the faradaic response I_F, due to the reduction or oxidation of respective bonds or substituents of the adsorbed material. As one is interested to obtain responses only from the adsorbed material one eliminates virtually contributions from further material diffusing from the bulk of the solution to the interface by applying very high sweep rates between 1 and 50 V s^{-1} for the applied electrode potential. Alternatively one can make use of the non-faradaic response of the charging current I_c being proportional to the differential double-layer capacitance C_D and therefore indicating adsorption, desorption and interfacial reorientation events. Usually both alternatives will be exploited by application of suitable voltammetric methods in investigations of this kind.

The block diagram given in Fig. 2 shows the basic components of the *voltammetric circuit*. Except the cell, all other components are parts of instruments termed polarographs or voltammetric analyzers. If modes of pulse voltammetry are applied and therefore pulsed polarization of the working electrode is required the components pulse generator and time generator have to be included into the operational components.

The voltammetric cell is equipped with three electrodes, the working electrode, an auxiliary electrode and a reference electrode, which is usually a saturated calomel electrode (s.c.e.) or an Ag/AgCl-electrode connected to the solution by a salt bridge. This 3-electrode technique permits potentiostatic control of the potential adjusted at the working electrode. This has the advantage that even for high ohmic resistance of the solution, due to low supporting electrolyte concentrations of 10^{-2} M or even 5×10^{-3} M, voltammograms can be still recorded. A suitable electronic potentiostat is incorporated into all modern polarographs.

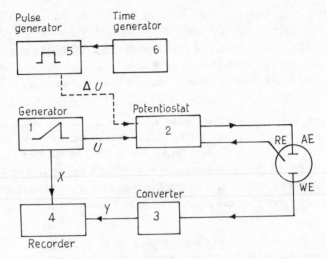

Fig. 2. – Block diagram of voltammetric circuit utilizing 3-electrode technique with potentiostatic potential control and optionally pulse polarization if pulse voltammetry is performed. (1) voltage generator with linear dc-ramp; (2) potentiostat; (3) current/voltage converter; (4) X-Y-recorder; (5) pulse generator for rectangular voltage pulses of height ΔU; (6) time generator for clock time of voltage pulses. Cell with WE working electrode, AE auxiliary electrode and RE reference electrode

Figure 3 gives a survey on the main features of the voltammetric methods most significant for applications in bioelectrochemistry. For more detailed and comprehensive information on the methodology, theory and general application areas of voltammetry reference is made to the books, monographs and articles cited in Ref. 4-18.

3.   Investigations of mono- and oligonucleotides

Among the nucleotides the adenine- and cytosine-nucleotides are reducible at the mercury electrode. They have particular biological significance not only as the building stones of nucleic acids, but also as monomers in various physiological processes.

The adenine phosphates 5'-AMP, 5'-ADP, 5'-ATP, 3'-AMP and cyclic 3', 5'-AMP are of significance for the carbohydrate cell metabolism and the concomitant function of organs, $e.g.$ muscle contraction, cardiovascular system, as well as for hormone production in the adrenal cortex. 5'-AMP and 3'-AMP are also of

Voltammetric Modes of Bioelectrochemical Significance

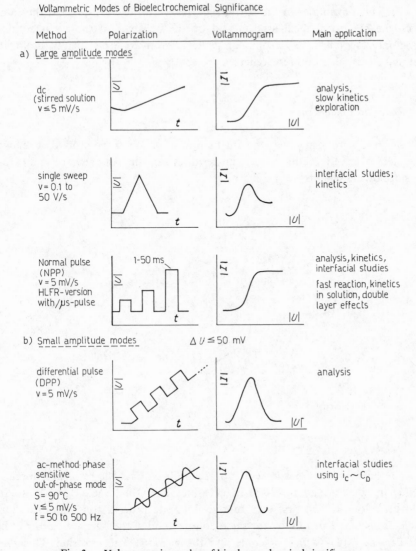

Fig. 3. — Voltammetric modes of bioelectrochemical significance

importance in the biotechnological process of alcoholic fermentation. Cyclic 3', 5'-AMP is a chemical messenger with key functions for the regulation of enzymes involved in the storage of sugar and fats in the living cell. 5'-AMP and 3'-AMP are furthermore building units of the nucleic acids. Oligonucleotides as CpC, *i.e.* Cytidyl-(3'-5')-cytidine, and ApA act as templates for DNA-polymerase in the production of double stranded biopolymers.

The given examples of biological significance make obvious, that the elucidation of the electrochemical aspects and of the behavior of those nucleotides at charged interfaces had to become of great interest to a number of biologically orientated electrochemical researchers.

3.1.   Reduction mechanism

Of great significance was the elucidation of the elctrode process of reduction of the electroactive moieties in these nucleotides, *i.e.* the bases adenine and cytosine [19] (s. following scheme):

The prerequisite for the reduction of both bases is the prior protonation of the N(1) in adenine and the N(3) in cytosine. Therefore, the pH of the solution should be below 7. The first step of the electrode process consists in the uptake of 2 electrons and 1 proton by the same double bond in the pyrimidine ring of both bases, *i.e.* the N(1)=C(6) in adenine and the N(3)=C(4) in cytosine. This step is rate-determining for the electrode process. Subsequently adenine takes up two further electrons and protons. Cytosine has to undergo a chemical deamination process before further electron and proton uptake is possible. As the deamination is slow the electron and proton uptake of cytosine implemented into oligo- and polynucleotides remains virtually restricted to $n = 2$ while for adenine $n = 4$. Only for rather long electrolysis times, not relevant for voltammetric conditions, a rather slow deamination of the reduction product can occur to a sufficient extent to enable an even further reduction by the uptake of further two electrons and protons ($n = 6$).

3.2. Characterization of interfacial behavior by a.c.-voltammetry

With respect to the frequent interactions of nucleotides with charged interfaces in living systems the voltammetric investigation of fundamental physicochemical aspects of their interfacial behavior at the model interface electrode/solution
has obtained particular significance. The results of two suitable voltammetric methods support each other. The first method is a low amplitude technique, *i.e. a.c.*-
voltammetry at the hanging mercury drop electrode (h.m.d.e.). The usage of this
stationary electrodes allows to adjust the adsorption time t_s at each potential value
to such time spans that adsorption equilibrium can be established according to the
respective bulk concentration of the nucleotide. As the interfacial situation is to
be studied use in made of the phase dependence of the *a.c.*-response. By adjusting
a phase angle of 90° between the polarizing *a.c.*-voltage with a peak-to-peak amplitude of 5 mV_{pp} and the resulting *a.c.*-current $I_{a.c.}$, its out-of-phase component
corresponding to the non-faradaic component of $I_{a.c.}$ is recorded. This non-faradaic
component of $I_{a.c.}$ is proportional to the differential double layer capacitance of the
electrode at the respective electrode potential. This voltammetric approach is
sufficiently precise and efficient but much more rapid than the in the past to analogous problems applied cumbersome and time consuming determination of the
differential double-layer capacitance by an impedance bridge.

Figures 4 to 6 show examples of such investigations [20] by phase sensitive
a.c.-voltammetry for 5'-AMP, 5'-ADP and 5'-ATP. At lower bulk concentrations of
the nucleotides progressive coverage by a *dilute* adsorption layer is indicated. But at
higher bulk concentrations around the potential of zero charge, $U_{e.c.m.} = -0.6$ V,
a pit occurs reflecting the formation of a *compact* film on the electrode surface.
The pit grows in depth and width with the bulk concentrations. After the pit at
−0.95 to −1.1 V a non-faradaic reorientation peak is observed which corresponds to
the reorientation of the adenine from a flat to a more perpendicular position in the
again dilute adsorption state. This reorientation peak has been observed also for all
adenine-containing oligo- and polynucleotides including DNA, and also for adenine
itself. A similar reorientation peak is also obtained for the cytosine nucleotides [21].

In this context it is emphasized that the extended investigations have furnished manifold evidence that the adsorption of mono- and polynucleotides always
occurs by their base moieties. This is to be expected as they are the hydrophobic
part of nucleotides while the sugar and usually also the phosphate being free ro-
tatable have with respect to their hydrophilic character a tendency to be orientated
towards the solution. In the slighly acidic milieu (pH 3,4) usually adjusted in our
investigations the N(1) in adenine and N(3) in cytosine are protonated and carry
thus a positive charge. Its effect is for 5'-phosphorilated nucleotides, however, fully
or largely compensated by the negative charge of the phosphate group particularly

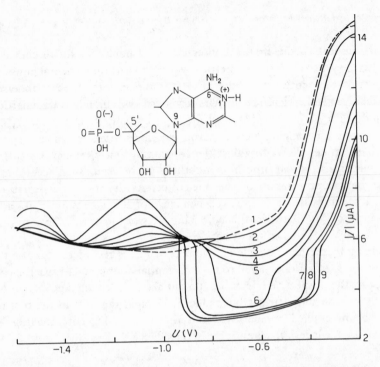

Fig. 4. — *a.c.* Voltammetric curve of adenosine-5'-monophosphate (5'-AMP) at the h.m.d.e. 0.5 *M* McIlvaine buffer, pH 3.4, 5 °C, area A of h.m.d.e. 3.5×10^{-2} cm^2, scan rate 5 mV s^{-1}, frequency 310 Hz, amplitude 5 mV$_{pp}$, phase angle 80°. Curve (1) 0; (2) 2.1×10^{-6}; (3) 5.4×10^{-6}; (4) 1.1×10^{-5}; (5) 4.3×10^{-5}; (6) 8.6×10^{-5}; (7) 2.1×10^{-4}; (8) 4.3×10^{-4}; (9) 8.6×10^{-4} *M* 5'-AMP

as the phosphate group orientates rather closely to the protonated N(1) or N(3) of the respective base type. Despite of the resulting overall zwitterionic character of the monophosphates there will be additional coulombic interactions according to the sign and amount of the charge of the interface. Thus, different types of orientation and adsorption will occur according to the potential of the electrode with respect to the zero charge potential ($U_{e.c.m.}$). For instance for adenine nucleotides at positive potentials the protonated N(1) will be repelled and adsorption will be mainly maintained by the attracted negatively charge phosphate. This interfacial interaction of phosphate in the appropriate potential range, providing only a dilute adsorption layer of the nucleotide irrespective of its bulk concentration, has been meanwhile also immediately confirmed for adenine nucleotides and DNA by Surface Enhanced Raman Scattering (SERS) [22]. For details of these studies see the respective section.

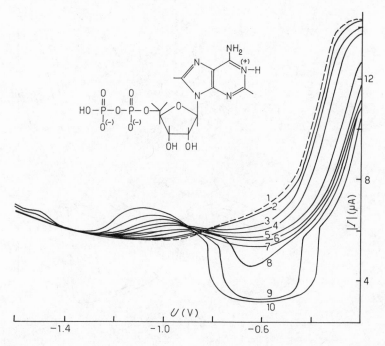

Fig. 5. — *a.c.* Voltammetric curve of adenosine-5'-diphosphate (5'-ADP) at the h.m.d.e. Curve (1) 0; (2) 1.0×10^{-6}; (3) $4.1. \times 10^{-6}$; (4) $7.5. \times 10^{-6}$; (5) 2.1×10^{-5}; (6) 4.1×10^{-5}; (7) 8.2×10^{-5}; (8) 1.0×10^{-4}; (9) 2.1×10^{-4}; (10) 4.1×10^{-4} M 5'-ADP

At potentials around $U_{e.c.m.}$ the adsorption will occur via the bases being the most hydrophobic part. In the dilute state a rather flat orientation prevails providing interaction of the π-electron system with the electrode surface. At elevated bulk concentrations a compact film forms as is reflected by the pit formation. The bases are now more vertically orientated with the phosphate, sugar and protonated N(1), *i.e.* the hydrophylic parts, pointing towards the solution.

At potentials several hundred mV more negative than $U_{e.c.m.}$, a reorientation of the adsorbed base moiety occurs. In the dilute state this reorientation, indicated by the mentioned peak between −0.9 and −1.1 V (s.c.e.), tilts the base from the flat to the perpendicular position with the protonate N(1) pointing towards the interface, due to coulombic attraction by the now negatively charged electrode surface. If there was a compact film this would collapse and the bases would be tilted as well into the same position. At even more negative potentials beyond −1.4 V (s.c.e.) the reduction of the adenine occurs. This is reflected in the out-of-phase *a.c.*-voltammogram (see Fig. 4) by a small peak for high bulk concentrations, because the

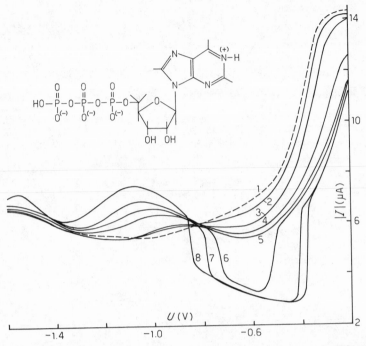

Fig. 6. — *a.c.* Voltammetric curve of adenosine-5'-triphosphate (5'-ATP) at the h.d.m.e. Curve (1) 0; (2) 2.1×10^{-6}; (3) 7.7×10^{-6}; (4) 2.1×10^{-5}; (5) 1.0×10^{-4}; (6) 2.1×10^{-4}; (7) 4.1×10^{-4}; (8) 8.2×10^{-4} *M* 5'-ATP

pseudo-capacitance connected with the electrode process causes even in the out-of-phase mode a certain response, although the current recorded in this mode is insensitive to faradaic events and is primarily of non-faradaic *i.e.* capacitive nature.

3.3. *Interfacial studies by single sweep voltammetry*

The foregoing findings and conclusions on the interfacial behavior of nucleotides at charged interfaces are based on the non-faradaic responses obtained by the out-of-phase mode of *a.c.*-voltammetry. An independent alternative voltammetric approach is single sweep voltammetry. Although also with this technique non-faradaic responses can be recorded, as has been shown for DNA, the completely independent alternative considered here relies on the faradaic response, due to the reduction of adenine present in the adsorbed amount of nucleotide [23]. Similar investigations have been performed with the cytosine nucleotides [21, 24, 25].

The point is to measure only the reduction response of the adsorbed nucleotide amount and to eliminate practically any contribution from dissolved material diffusing towards the interface. This can be achieved by applying high sweep rates ($\leqslant$ 20 V s^{-1}), moderate bulk concentrations (10^{-5} M for adenine- and 10^{-4} M for the less surface active cytosine-mononucleotides) and sufficiently long adsorption times t_s of 60 to 240 s, respectively, at the sweep starting potential U_s to achieve a steady state of surface coverage before the sweep is applied. During the rapid voltage sweep to the reduction potential range of the base moiety (adenine or cytosine) of the nucleotide, the adsorption state, and interfacial orientation, of the adsorbed nucleotide (adjusting itself during t_s at the respective sweep starting potential value U_s) is practically not altered. As due to the high sweep rate also diffusion contributions remain negligible, the faradaic reduction response of the adsorbed nucleotide can be used in this manner as a valuable sensor to study its interfacial behavior. Utilizing the integration circuit incorporated in the modern polarographs, the total amount of the adsorbed nucleotide is recorded immediately by registrating the charge Q required for its reduction. Q corresponds to the interfacial concentration Γ of the adsorbed nucleotide according to equation (5).

$$Q = nFA\Gamma \tag{5}$$

where n is the number of electrons involved in the reduction of each mononucleotide and equals to 4 for adenine nucleotides and 2 for cytosine nucleotides, F is the Faraday constant, A the electrode surface in cm^2 and Γ the surface concentration of the nucleotide in mole cm^{-2}. The amount of adsorbed nucleotide, expressed as Q or Γ will depend on the bulk concentration c and the adsorption potential U_s which equals for sweep voltammetry to the respective value of the sweep starting potential. Figure 7 shows the dependence of the adsorbed amount Q of various adenine nucleotides and the nucleoside adenosine as function of the adsorption potential U_s for a bulk concentration of 2 x 10^{-5} M and an adsorption time t_s sufficiently long to attain adsorption equilibrium. The bulk concentration is below the threshold value where compact film formation becomes possible and thus the curves in Fig. 7 refer to the dilute adsorption state. It should be noted in this context that in the dilute and compact state always monomolecular adsorption layers are formed. The analysis of the functions in Fig. 7 lead to the same conclusions as they emerge from the out-of-phase a.c.-voltammograms. The relatively strongest adsorption is observed for the relatively most hydrophobic species the nucleoside adenosine. In the potential range around $U_{e.c.m.}$ the introduction of further hydrophilic phosphate groups lowers significantly the adsorption degree for 5'-ADP and 5'-ATP. Also the increasing repulsion of the nucleotides at potentials more negative than the $U_{e.c.m.}$-range is clearly reflected while with increasing number of phosphates the adsorption range is somewhat more extended at potentials significantly more positive than $U_{e.c.m.}$. The supporting effect of the uncompensated

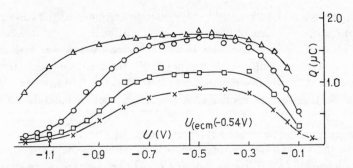

Fig. 7. – Dependence of the charge Q due to reduction of adsorbed species on the potential of adsorption. 0.5 M McILVAINE buffer, pH 3.4, 5 °C, time of adsorption 60 s, v 20 V s^{-1}, c 2.0 x 10^{-5} M, A 5.0 x 10^{-2} cm^{-2}. Curve (1) adenosine, (2) 5'-AMP, (3) 5'-ADP, (4) 5'-ATP

positive charge on the protonated N(1) in the nucleoside adenosine is responsible for the extension of substantial adsorption to the potential range considerably more negative than $U_{e.c.m.}$ [23, 26].

3.4.   Adsorption isotherms

The quantitative evaluation of the measurements by out-of-phase *a.c.*-voltammetry or the single sweep method as function of the bulk concentration yields the adsorption isotherm at the respective adsorption potential [26].

The adsorption isotherms, Fig. 8 shows an example, are of the FRUMKIN type, due to the lateral interactions between the adsorbed nucleotide or nucleoside particles. In the pit region a double step isotherm is obtained with a first plateau corresponding to full coverage in the *dilute* and a second plateau for saturation in the *compact* adsorption stage.

The equation of the FRUMKIN-isotherm is

$$bc = \frac{\Gamma/\Gamma_m}{1 - \Gamma/\Gamma_m} \exp\left(-2\,a\,\Gamma/\Gamma_m\right) \tag{6}$$

where b is the adsorption coefficient in l mole^{-1}. As it is related to the free adsorption enthalpy according to:

$$\Delta G_{ad} = RT \ln(55.5\, b) \tag{7}$$

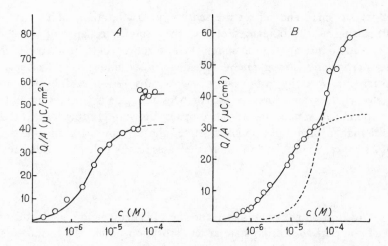

Fig. 8 *A*. — Half logarithmic plot of adsorption isotherms for 5'-AMP from phase sensitive *a.c.*-voltammetric and sweep voltammetric measurements. Adsorption potential U_s −0.6 V; pH 3.4; 0.5 *M* McILVAINE buffer.
B. — Half logarithmic plot of adsorption isotherm for 5'-ADP. Experimental curve (1), theoretical fitting for the first step (3) and for the second step (2).

it is a measure of the adsorption tendencies. *b* is a function of the adsorption potential U_s, *c* is the bulk concentration in moles l^{-1} and *a* the lateral interaction coefficient having positive values for attraction.

Table 2 summarizes the results for the adenine mononucleotides at the potential of zero charge $U_{e.c.m.}$ for the *dilute* and *compact* adsorption state at full coverage of the interface with a monomolecular layer, *i.e.* at Γ_m.

From Tab. 2 the following tendencies emerge for the *dilute* adsorption state.

Γ_m decreases by 40 % from the nucleotide adenosine to 5'-ATP, the adsorption coefficient *b* decreases by a factor 2 and the adsorption free enthalpy $\Delta G°$ by 1.6 kJ mole^{-1}. The interaction coefficient *a* decreases to zero with increasing number of phosphate groups, due to increasing lateral repulsion of the negatively charged 5'-ADP and 5'-ATP. At the same time the adsorption area S_m required per adsorbed nucleotide species increases. Restricted possibilities for charge compensation in the protonated 3'-AMP are the reason for the relatively large S_m-value and the compared with 5'-AMP significantly smaller *b*-value.

For the *compact* film state, corresponding to the second step in the FRUMKIN-isotherm, rather equal Γ_m-values are found, except for adenosine being a positively charged species, due to N(1) protonation uncompensated by phosphate.

The *b*-values are small and reflect the sensitivity to alterations of the adsorption potential causing at the boudaries of the pit a sudden collapse of the compact film back to the dilute adsorption state. The interaction coefficient *a* is rather high, indicating particularly strong *lateral* stacking forces among adsorbed species in the compact film, specially in the case of zwitterions 5'-AMP and 3'-AMP in which the selective charges are more or less compensated. The S_m-values are still large enough to permit reorientations in the compact film state. In this context it should be stressed that in oligo- and polynucleotides the S_m-values for adenine are by a factor of 2 smaller, due to the fixation of the bases on the polynucleotide backbone.

Table 2. — Adsorption parameters of adenosine and adenine mononucleotides in the diluted and compact adsorption state at an adsorption potential of −0.6 V (s.c.e.); 5 °C; pH 3.4; h.m.d.e. according to Ref. 23.

Substance	$10^{10}\,\Gamma_m$ (mole cm²)	S_m (Å²)	$10^5\,b$ (l mole⁻¹)	$-\Delta G°$ (kJ mole⁻¹)	a
Dilute state					
adenosine	1.11	150	1.93	37.7	0.45
5'-AMP	1.09	153	2.39	38.2	0.40
5'-ADP	0.90	185	1.50	37.1	0.40
5'-ATP	0.73	229	0.96	36.1	0.00
3'-AMP[*]	0.52	317	1.48	37.1	0.45
3', 5'-AMP	0.77	216	2.34	38.1	0.40
Compact state					
adenosine	2.07	80	0.040	28.9	1
5'-AMP	1.4	118	0.001	20.5	5
5'-ADP	1.6	103	0.067	29.7	1
5'-ATP	1.5	112	0.019	26.8	1
3'-AMP[*]	1.6	103	0.002	22.2	5

[*] at −0.8 V (s.c.e.)

In general cytosine nucleotides show an analogous interfacial behavior [21]. The surface activity of pyrimidine bases is, however, smaller than that of purine bases as consequence of the different dipole moment. But the stacking interaction in the compact film state is stronger than for the corresponding adenine mono-nucleotides [25].

4.   Applications to the biophysical chemistry of nucleic acids

As carrier of the genetic code deoxyribonucleic acid (DNA) has a key function in all living systems. Interactions with charged biological interfaces, *e.g.* enveloping proteins, with enzymes or intercellular membranes, are frequent and have important consequences. An example is the initiation of strand separation for DNA replication in the course of cell mitosis. The application of suitable voltammetric modes, partially optimally tailored for the specific application, have contributed significantly to the elucidation and understanding of general bioelectrochemical parameters determining the behavior at charged interfaces of native double helical DNA in comparison with that of denatured single stranded DNA [30, 31]. In this context it has to be emphasized that the results of investigations on the interfacial and electrochemical behavior of the building stones of the biopolymer DNA, *i.e.* the mono- and oligonucleotides [19-28] of adenine, cytosine and guanine as well as the separate bases and in addition also biosynthetic hetero-oligonucleotides and homo-polynucleotides, *e.g.* poly-A[29], have furnished a variety of supporting results and evidence necessary to understand and interpret correctly the measurements on the complex interfacial behavior of DNA. Moreover, the information potentialities provided by voltammetry have been extended significantly by the fact that differences in the percentage of the base composition of DNA from different origin are reflected in the experimental data. In this manner the general contours of the interfacial behavior of native DNA could be understood and modelled from appropriate systematic voltammetric measurements at the easily controllable model interface electrode/solution. Recent applications of SERS have furnished by the obtained surface enhanced Raman spectra further confirming evidence by an independent approach [32]. It is to be emphasized that the elucidation of the subject has reached a state that it is now even possible to fit the major results obtained in the past with other less informative voltammetric and polarographic methods into the general picture which has emerged. Based on these fundamental results and perceptions appropriate voltammetric studies have revealed new findings on hitherto not known or not clarified damages induced in native DNA by ionizing radiation or by mutagenic chemicals and have provided the basis for the elaboration of simple and reliable voltammetric *in vitro* test procedures to monitor the damage potential of radiation or certain types of mutagenic chemicals on different types of native DNA. Such *in vitro* tests will reflect the full damage potential as there occurs no antagonistic action by repair enzymes operative in common biological test procedures, *e.g.* the AMES test.

4.1.   Biophysicochemical aspect of the behavior of DNA at charged interfaces

A decisive strategical point for the investigations on the interfacial behavior of DNA was the selection of an optimally appropriate voltammetric method from the arsenal of available possibilities. A basic requirement was that the chosen method should be focussed on the amount of DNA being adsorbed at the interface and contributions by the bulk concentration of DNA in the solution should be excluded. Therefore single sweep voltammetry with high sweep rates (0.1 to 1 V s^{-1}) at a stationary mercury electrode, the h.m.d.e., was chosen as the main method. This approach enables the investigator to adjust the adsorption time t_s at a chosen potential U_s at will. In this manner not only adsorption equilibrium can be attained before measurement but during t_s also the subsequent interfacial events of adsorbed DNA can reach a steady state corresponding to the respective adsorption potential U_s adjusted and the concomitant interfacial electric field. Subsequently the potential is driven by a rapid voltage sweep (see Fig. 9) into the potential range of the reduction of adenine and cytosine and beyond that until the onset of hydrogen evolution. Both bases are among the four base types (adenine, cytosine, guanine, thymine) contained in DNA, those which are reducible under certain preconditions at pH values below 7. Therefore most measurements have been made in 0.5 M McILVAINE buffer at pH 5.6 with additions of KCl to adjust the ionic strength.

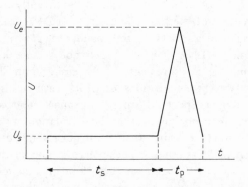

Fig. 9. – Principle of polarization by single triangle sweep voltammetry. U_s sweep starting potential and adsorption potential. U_e potential of sweep reversal; t_s adsorption time elapsed at U_s; t_p duration of triangle sweep

Figure 10 shows an example for the voltammograms obtained with single stranded (s.s.) denatured DNA. This is, of course, compared with native double helical DNA the much more simple case, because s.s.-DNA adsorbs immediately at the interface *via* its base units and the reducible base types adenine and cytosine are at the reduction potential immediately accessible to electron and proton transfer in the course of the electrode process. The electrode process of the reduction of adenine and cytosine has been treated in the section on nucleotides. It is emphasized already here, that the electrochemical reduction of the reducible bases, adenine and

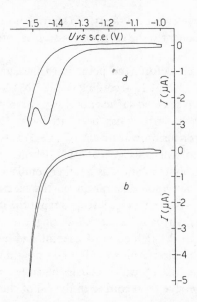

Fig. 10. — Sweep voltammetric current-potential curve of adsorbed denatured DNA. 0.5 M McILVAINE buffer, pH 5.8, 240 μg/cm^3 denatured DNA; sweep rate 100 mV s^{-1}; t_s 20 s at U_s of -1.0 V. (*a*) first curve fresh mercury drop showing reduction response; (*b*) second curve at same drop after waiting 15 min at -1.0 V showing no reduction response

cytosine, of adsorbed DNA is only possible, if both steps of the electrode process, electron and proton uptake, can occur. This requires that the acceptor sites for proton uptake at the adenine and cytosine moieties are available and not occupied by interstrand hydrogen bonding as in intact double helical native DNA.

The voltammogram in Fig. 10 reflects, that it can be only obtained at a fresh mercury drop covered with adsorbed DNA. Once the reduction of the adenine and cytosine has been performed by the first sweep no further electrode reaction is possible with repeated sweeps, because a dense film of highly surface active reduced DNA covers the electrode. This very stable film will not desorb and thus no fresh unreduced DNA can adsorb at the interface. Consequently subsequent sweeps will yield no voltammetric response except the base current and at potentials more negative than -1.7 V (s.c.e.) the hydrogen evolution response, which occurs at lower overvoltage than usual at mercury, because it is catalyzed by the adsorbed N-heterocycles, contained in reduced DNA, according to a mechanism elucidated over 20 years ago for this type of catalysts [33]. If, however, a fresh mercury drop is formed and the whole procedure is repeated under the same conditions (same t_s, U_s etc.) the first sweep at the fresh drop will reproduce the voltammogram with the reduction response. The form of the voltammetric curve (Fig. 10) indicates also that by returning with the voltage sweep to the sweep starting potential (Fig. 9) no reoxidation is achievable. Consequently the electrode process is totally irreversible.

As has been stated before, the point is to measure only the voltammetric response of the adsorbed DNA. If according to the DNA bulk concentration the adsorption time t_s has been selected sufficiently long, *i.e.* in practice 15 to 1000 s, full coverage of the electrode surface has been attained. The adsorption process is diffusion controlled. A diffusion coefficient of $(4 - 5) \times 10^{-7}$ cm^2 s^{-1} is determined for s.s.-DNA with a molecular weight of 10^6 daltons indicating that denatured DNA exists in the electrolyte solution as a rather compact and spherical coil, which is more or less spread out upon adsorption on the electrode surface, where it is adsorbed *via* base units. As the voltage sweep is rapid the interfacial situation of the adsorbed DNA remains virtually unaltered during the sweep and therefore the faradaic responses, due to the reduction of adenine and cytosine, can be utilized as a sensor for assessing the interfacial situation adjusting during the adsorption time t_s at the adsorption potential U_s which equals the sweep starting potential. Instead of the current it is preferable to record with the aid of the integration circuit in the polarograph the charge Q consumed for the reduction of adenine and cytosine of the adsorbed DNA [34].

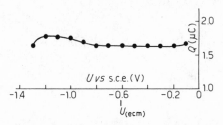

Fig. 11. — Dependence of sweep voltammetric reduction response (charge Q) on adsorption potential U_s at full coverage. 107 μg/cm^3 denatured DNA; BRITTON-ROBINSON buffer + 1 M KCl; pH 6.68; 25 °C; sweep rate 100 mV s^{-1}; t_s 30 s

The dependence of this charge Q on the adsorption potential U_s shows for s.s.-DNA a rather constant behavior over the whole U_s-range (Fig. 11). The slight enhancement at more negative U_s-values is connected with the decrease in required adsorption area S_m per adsorbed base unit, due the reorientation of the bases to a more perpendicular orientation in this U_s-range. This reorientation is shown for adenine units in Fig. 12 by supporting measurements with phase sensitive *a.c.*-voltammetry in the out-of-phase mode. A similar reorientation response has been observed for adenine mono- and oligonucleotides (see respective section).

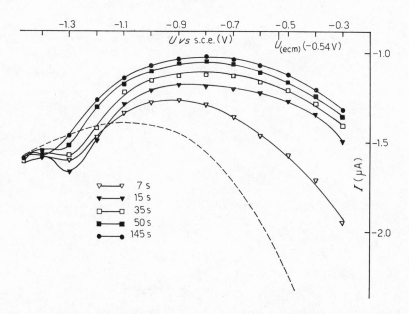

Fig. 12. — Out-of-phase component of *a.c.*-voltammetric current as function of adsorption potential U_s and adsorption time t_s for denatured calf thymus DNA (50 $\mu g/cm^3$); 0.1 *M* McILVAINE buffer, pH 5.6; 78 Hz; *a.c.*-amplitude 5 mV$_{pp}$, phase angle 90°; adsorption times t_s indicated. Broken line, buffer without DNA

Adsorbed native double helical DNA yields a similar reduction response as s.s.-DNA. However, a sequence of certain interfacial events, of which deconformation is of central significance, has to occur before during the adsorption time t_s elapsed at the respective U_s-value. This is reflected by the U_s-dependence of the faradaic response Q for adsorbed native and s.s.-DNA in Fig. 13. Only a fraction of the adenine and cytosine units are reducible in native DNA. This reducible amount is smallest if U_s coincides with the zero charge potential $U_{e.c.m.}$ and increases for U_s-values on both sides of $U_{e.c.m.}$, *i.e.* with increasing interfacial electric field. At a U_s of −1.25 V close before the onset of the reduction potential range Q-values for native DNA reach 80 to 90 % of those obtained over the whole U_s-range for s.s.-DNA. This behavior shows that adenine and cytosine units can be only reduced in those portions of adsorbed native DNA where interfacial helix opening has occurred [30, 31, 35-38]. This helix opening of the adsorbed regions of native DNA occurring possibly *via* the transitory stage of a linear ladder structure [67] is depicted schematically in Fig. 14. Thus, interfacial deconformation is a prerequisite for the reducibility of adenine and cytosine units in native DNA. Otherwise the proton uptake involved in the electrochemical reduction of the base moieties could not

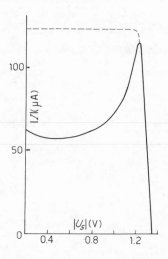

Fig. 13. — Dependence of sweep voltammetric reduction response I on adsorption potential U_s at full coverage for denatured DNA (---) and native calf thymus DNA (——)

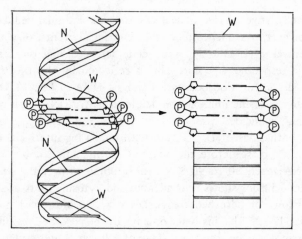

Fig. 14. — Scheme of transformation of double helix of native DNA into double stranded linear *ladder* structure

occur, because in the intact helix the proton acceptor sites at N(1) in adenine and N(3) in cytosine are occupied by interstrand WATSON-CRICK hydrogen bonds. Theoretical speculations in the literature which take not into account these basic requirements and ignore the existing evidence lead inevitably to irrelevant conclusions on the fate of native DNA at charged interfaces. The U_s-dependence of the reduction response Q in Fig. 13 reflects for adsorbed native DNA the degree of interfacial helix opening at the respective U_s and consequently for the corresponding interfacial electric field. This is a finding of fundamental significance.

The flow chart in Fig. 15 summarizes the different sequences of interfacial events for native double helical and denatured s.s.-DNA after adsorption. The destabilization of adsorbed native DNA is induced by the consequences of local dehydration of the adsorbed regions [36]. One consequence of the resulting destabilization is also a certain spread of adsorption along the helix according to the respective U_s-value.

Further evidence for the interfacial helix opening of adsorbed native DNA manifesting itself in the Q-U_s-dependence has been furnished also by investigations on the non-faradaic responses connected with reorientations of adenine and cytosine moieties no more bound by interstrand hydrogen bridges [39]. They yield a similar Q-U_s-dependence as the faradaic response connected with the reduction.

Also measurements with a specially tailored potentiostatic double-sweep-method [40] confirm the interfacial deconformation of DNA. Moreover, the Raman spectra recorded recently by the SERS-approach provide an independently obtained confirmation [32]. A further important independent confirmation for interfacial helix opening are· the in a subsequent section in more detail treated findings with native DNA which had been subjected before to ionizing irradiation.

It should be also noted that the for the first time by the mentioned voltammetric measurements experimentally observed interfacial helix opening is to be theoretically expected at charged interfaces for electric field strength values above 10^4 V cm^{-1} as they are certainly operative at the used model interface electrode solution [41]. Thus, it is to be concluded that the described voltammetric experiments model the fate native DNA experiences in biological systems upon interaction with charged biological interfaces, e.g. in the primary stage of cell mitosis, where separation of the DNA-strands is a prerequisite for later DNA-replication necessary to ensure the transfer of the genetic information.

Voltammetric investigations of the described kind provide also information on the effects of the base composition of DNA with respect to the stability of the double helical conformation. If one compares the Q versus U_s-dependencies for

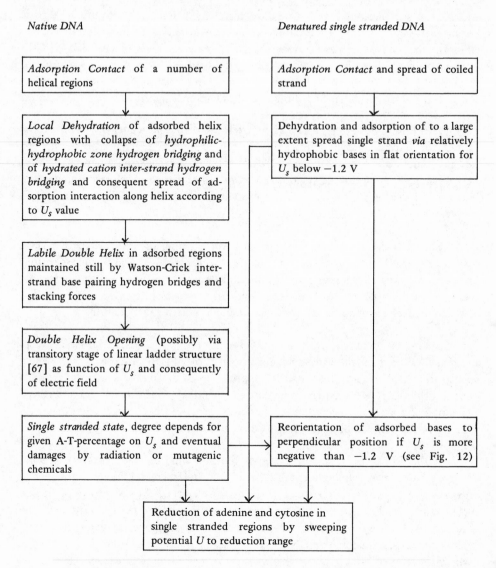

Native DNA

Denatured single stranded DNA

Adsorption Contact of a number of helical regions

Local Dehydration of adsorbed helix regions with collapse of *hydrophilic-hydrophobic zone hydrogen bridging* and of *hydrated cation inter-strand hydrogen bridging* and consequent spread of adsorption interaction along helix according to U_s value

Labile Double Helix in adsorbed regions maintained still by Watson-Crick inter-strand base pairing hydrogen bridges and stacking forces

Double Helix Opening (possibly via transitory stage of linear ladder structure [67] as function of U_s and consequently of electric field

Single stranded state, degree depends for given A-T-percentage on U_s and eventual damages by radiation or mutagenic chemicals

Adsorption Contact and spread of coiled strand

Dehydration and adsorption of to a large extent spread single strand *via* relatively hydrophobic bases in flat orientation for U_s below −1.2 V

Reorientation of adsorbed bases to perpendicular position if U_s is more negative than −1.2 V (see Fig. 12)

Reduction of adenine and cytosine in single stranded regions by sweeping potential U to reduction range

Fig. 15. — Flow chart of sequence of interfacial events of native and denatured DNA at charged interfaces during adsorption time t_s at adsorption potential U_s and concomitant interfacial electric field

calf thymus DNA and bacterial DNA from *Micrococcus Lysodeikticus* significant differences in the degree of helix opening are observed. This is due to the different strength of A-T and G-C-pairing by interstrand hydrogen bonding. The G-C-bonds are stronger than the A-T-bonds. Thus, a native DNA containing a higher percentage of A-T-bonds, *e.g.* CT-DNA, will yield under equal conditions a higher degree of interfacial helix opening than a G-C rich DNA as the bacterial ML-DNA. As Fig. 16 shows a linear relationship is obtained for various DNA-species and synthetic helical polynucleotides, if the faradaic response Q corresponding to a certain U_s and t_s and consequently degree of interfacial deconformation is plotted *versus* the adenine percentage [42]. Also differences in the molecular weight play a role for the resulting degree of helix opening as the different Q-values in Fig. 16 for DNA species

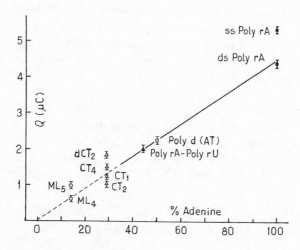

Fig. 16. — Dependence of sweep voltammetric reduction response (charge Q) of various double stranded polynucleotides on adenine content. 0.5 M McILVAINE buffer; pH 5.6; sweep rate 1 V s^{-1}; U_s 0.4 V.

ds-poly rA	25 μg/cm^3, M.w. 0.2 x 10^6
ss-poly rA	25 μg/cm^3, M.w. 0.2 x 10^6
poly d(AT)	35 μg/cm^3, M.w. 0.9 x 10^6
poly rU · poly rA	100 μg/cm^3, M.w. 1.0 x 10^6
calf thymus CT$_4$ DNA	870 μg/cm^3, M.w. 0.5 x 10^6
CT$_1$ DNA	85 μg/cm^3, M.w. 3.0 x 10^6
CT$_2$ DNA	150 μg/cm^3, M.w. 5.3 x 10^6
(denatured) dCT$_2$ DNA	50 μg/cm^3, M.w. —
Micrococcus Lysodeikticus	
ML$_4$ DNA	150 μg/cm^3, M.w. 9.0 x 10^6
ML$_5$ DNA	50 μg/cm^3, M.w. 0.8 x 10^6

with the same adenine content show. This is connected with differences in sterical hindrance for unwinding according to the fact, whether the double helical DNA to be adsorbed prevails in the more favourable rod like form in solution or in the more looped adsorption causing coiled form, which exists preferentially at molecular weights above 10^5 daltons [42].

4.2. Spectroelectrochemical Raman measurements with the SERS-method

Rapidly expanding significance for elucidating investigations on the interfacial behavior of nucleic acids and other biopolymers gains a new spectroelectrochemical approach termed Surface Enhanced Raman Scattering (SERS) [22, 43]. Some years ago it had been shown that pyridine adsorbed on a silver electrode gave Raman spectra with an at least 10^5 times enhanced intensity. Meanwhile we were the first to extend these SERS-measurements to the systematic investigation of adsorbed nucleotides [22, 43]. The characteristic bands have practically identical frequencies, as they are observed in normal Raman spectra of dissolved species in the bulk of the solution, but the sensitivity of SERS is several orders of magnitude higher. The SERS-spectra of the adenine mononucleotides yield results which confirm the conclusions drawn on their interfacial behavior from voltammetric studies.

Very recently we could go a step further to biopolymers and record for the first time the SERS-spectra of denatured single stranded and native double helical calf thymus DNA adsorbed at the silver electrode (Fig. 17) [32]. The vibration bands corresponding to the adenine moieties show a specific enhancement while

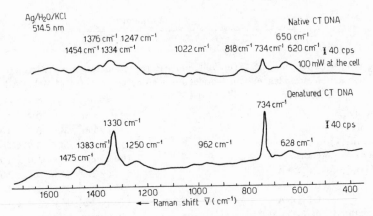

Fig. 17. — SERS-spectra of native and denatured CT-DNA. 0.1 M KCl + 2 x 10^{-3} M Na$_2$HPO$_4$; 225 μg DNA/cm^3; pH 8.0. Laser power at electrode 100 W. Prior activation of silver electrode by two triangular voltage sweeps between −0.1 and +0.2 V at a sweep rate of 50 mV s^{-1}

the other bases produce only relatively weak SERS-signals. The characteristic adenine bands occur for denatured and native DNA but with a lower intensity in the latter case, because the helix is opened only to a certain extent at the adjusted potential and the amount of with the surface interacting adenine moieties is consequently less than for single stranded denatured DNA. Moreover, the stretching frequency of the polymer backbone at 818 cm^{-1} observed for native DNA only, indicates that the helical structure remains intact to a certain degree.

Determinations of the kinetic parameters of hydrogen-deuterium exchange as function of temperature will provide a useful tool to study structural fluctuations of DNA, *i.e.* breathing, outside the melting temperature range [44].

4.3. Investigations of damages in native DNA by ionizing radiation and mutagenic chemicals

It has been well known that ionizing radiation, *e.g.* γ-radiation, causes damages in DNA, *via* radicals formed in radiation chemical reactions with the solvent water. These since many years in radiation biology well known kinds of damages are single strand breaks and at higher doses even an increasing number of double strand breaks followed by the release of oligonucleotide units.

Voltammetric investigations by the described single sweep method have revealed a further latent damage occurring already at rather low radiation doses, between about 0.1 and several krad [45, 46]. This newly detected damage consists in the destabilization of a number of base pairs and manifests itself upon adsorption at a charged interface by the corresponding Q *versus* U_s-relationship (Fig. 18). The higher the radiation dose, the higher becomes the degree of helix opening for the adsorbed irradiated native DNA. By comparison with the Q *versus* U_s-relationship referring to non-irradiated native and denatured s.s.-DNA the additional destabilization caused by the radiation can be assessed. Up to 5 krad a linear dose-effect relationship results. Comparing the efficiency of the in this manner detected helix destabilization with that of the local damages produced by single strand breaks it turns out, that per single strand break 100 to 150 base pairs are destabilized in the considered low γ-ray dose regime up to 4 krad. In contrast to the localized formation of single strand breaks there is no dependence on the ionic strength observed for the radiation induced helix destabilization. This new finding on hitherto unknown damaging effects by low radiation doses is of fundamental importance for radiation biology. While the destabilizations remain irreversible *in vitro*, it is to be expected, that they are largely though not completely repaired *in vivo*. The *in vivo* remaining small extent of radiation damage could play a key role for the correct enzymatic recognition of the secondary and tertiary structure of native DNA.

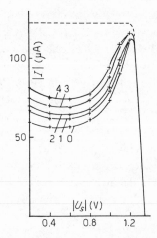

Fig. 18. — Dependence of sweep voltammetric reduction response I on adsoprtion potential U_s at full coverage for denatured, native and γ-irradiated native DNA, 75 μg/cm^3, ionic strength 0.0075 M, t_s 150 s, 0.5 M McILVAINE buffer, pH 5.6, sweep rate 10 V s^{-1}, 0: 0; 1: 0.88; 2: 1.86; 3: 2.84; 4: 3.82 krad; dotted line 5: denatured DNA

Besides the fundamental significance of the described findings emerging from the application of an appropriate voltammetric method its sensitivity and reliability provides the possibility for the introduction of a simple *in vitro* test procedure on the effect of small radiation doses in native DNA as function of its A-T percentage and consequently its origin. Such a test procedure is at present under development in our institute.

In this context it is mentioned that a similar voltammetric approach has also revealed the mechanism of alkylating substances causing *in vitro* mutagenic chemical damages in native DNA [47]. Methylation occurs preferentially, though not exclusively, at the N(7) position of guanine labilizing the affected G-C-pairs. This is as well reflected by the Q *versus* U_s-curves of methylated DNA, yielding also a linear dose-effect relationship as function of the methylation degree for the amount of additionally induced helix opening after adsorption of the methylated DNA.

The conversion of guanine into 7-alkylguanine can be regarded as a measure of the alkylation degree of nucleic acids. This alkylation degree can be determined by a new voltammetric procedure superior to existing non-electrochemical alternatives with respect to sensitivity, rapidity, reliability and convenience [48].

A 1 cm³ sample of the alkylated nucleic acid is first subjected to rapid hydrolysis at 100 °C by 1 *M* HClO₄ lasting 15 min. Then a pH 2 is adjusted with Mc-
ILVAINE buffer and an anodic differential pulse voltammogram is recorded at a
glassy carbon electrode. The oxidation peaks of guanine, 7-methylguanine (and
adenine) appear at well separated potentials (Fig. 19). From the ratio of the peaks
for guanine and 7-methylguanine follows the methylation, or more generally the
alkylation, degree of the tested nucleic acid.

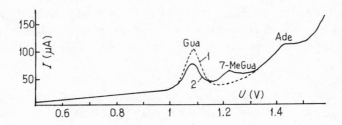

Fig. 19. — Differential pulse voltammogram of the oxidation of the acid hydro-
lysate of ML-DNA at Ph 1.2; (1) ML-DNA, (2) methylated ML-DNA with 14.4 %
7-MeGua

The oxidation of guanine has been used recently by other authors as well
for studies on DNA with differential pulse voltammetry at the graphite electrode
and cyclic sweep voltammetry at the dropping mercury electrode [49, 50]. A comprehensive review on the application of differential pulse voltammetry at carbon
electrodes for manifold fundamental investigations on DNA and polynucleotides
by responses due to electro-oxidation of base units has been recently published
[51].

4.4. *Interactions of DNA with enzymes and drugs*

This subject is an important further application of voltammetry in nucleic
acid chemistry.

An example is the proof that the electrochemical oxidation and enzymatic
oxidation with peroxidase of 3.9-dimethyl uric acid proceeds *via* an identical reaction mechanism as has been clarified by voltammetric measurements [52]. The
extension of these studies to uric acid, other N-methylated uric acids, uric acid-9-
riboside, guanine and 8-oxy-guanine has shown that this is also the case for these
substances and that voltammetric studies can provide uniquely valuable informations
on the complex mechanisms of enzymatic oxidations [53].

The application of *a.c.*-polarography has demonstrated the superiority of the voltammetric approach to follow the enzymatic degradation of mammalian and bacterial DNA by endonucleases (DNAse I, II) [54].

The degradation of calf thymus DNA with exonuclease III of *E.coli* occurs by specific cleavage of nucleotides from the 3'-ends while in nicked circular DNA intramolecular single strand regions are created. Studies with linear sweep voltammetry suggest that double helical DNA with single stranded ends is fixed by these in such a manner at the interface that the double stranded regions cannot come into contact with the interface and consequently helix opening by the action of the interfacial forces and electric field seems to be largely inhibited in this special case in contrast to the usual interfacial behavior of native DNA reported in the previous section [55].

The binding of anthracyclines, acting as antitumoral drugs, with DNA has been followed by voltammetry of the free anthracycline utilizing their 2-3 orders of magnitude higher diffusion coefficient compared with that of the DNA-drug complex [56]. In further detailed investigations four groups of anthracyclines could be distinguished. Low binding with DNA show derivatives with neutral sugar (*e.g.* steffimycin), medium binding is observed for derivates with one basic sugar (*e.g.* daunomycin, adriamycin, iremycin, carminomycin, 1-deoxypyrromycin), more pronounced binding is the case with derivatives containing one neutral and one or more basic sugars (aclacinomycin, marcellomycin) and highest binding exert the violamycines containing two basic sugars [57].

5.   Applications in food control, toxicology and ecotoxicology

5.1.   *Analysis by high performance liquid chromatography with voltammetric detection*

There takes place a growing application of voltammetric methods to the determination of organic substances of biological significance. Frequently with organic substances the problem has to be overcome that the substances to be studied have to be separated from others producing voltammetric responses in practically the same potential range. This need has been the reason for the steeply expanding application of the combination of an efficient separation procedure with a superior determination approach resulting in high performance liquid chromatography with voltammetric detection (HPLC-VD) [58].

The introduction of a voltammetric detector in HPLC provides not in all cases a substantially superior performance compared to the in the past common UV-detection but is doubtless significantly superior for a growing number of certain substance classes. They are at present predominantly those kinds of substances for which the electrode process of oxidation at a carbon paste, graphite or glassy carbon working electrode can be used while for substances to be determined by their reduction at mercury electrodes, HPLC with voltammetric detection frequently shows not yet advantages compared with UV-detection. The reason is connected with the principle of voltammetric detection usually to be applied in connection with prior HPLC-separation. Commonly a stream or jet of eluate flows from the column onto the surface of the, therefore preferentially solid, working electrode kept at an appropriately adjusted potential. The resulting current, due to the oxidation of the respective substance in the respective eluate fraction is determined as a function of eluation time and consequently the HPLC-chromatogram of the studied substance mixture is obtained.

The recorded current I depends in the following manner on the volume flow rate F_r and the substance concentration c:

$$I = k \, n \, \mathsf{F} \, c \, F_r^{\alpha} \tag{8}$$

where k is a constant depending on the kinematic viscosity, the diffusion coefficient, the electrode area and the cell geometry. The exponent α has in flowing streams values between $1/3$ and $1/2$, but in wall-jet cells it can reach $3/4$. Mostly the voltammetric detector is operated in the $d.c.$-mode to avoid sloping of the background current. Only in the particular case of strong adsorption of the oxidation products at the electrode surface the utilization of normal pulse voltammetry becomes recommendable and sometimes additional electrochemical separation power may be gained by the application of differential pulse voltammetry. In general it can be stated that for the growing number of substance classes, where the HPLC-VD approach is applicable, it is competitive in sensitivity with non-electrochemical alternatives as gas chromatography or spectrophotometry and spectrofluorimetry.

The potentialities of HPLC-VD are well featured by recent studies on amino compounds utilized as drugs and phenolic substances of high biological and ecological significance, because they are used as growth-promoting anabolic substances in cattle growing or occur as mycotoxins in cereal products. Fig. 20 shows the simultaneous trace determination of seven estrogen derivates and zearalenone in chicken meat [59, 60]. The determination sensitivity is by a factor 100 higher than with the UV-detector and reaches 0.5 ng/g or 0.5 ppb. The method can be used also advantageously for monitoring the estrogenic growth promoting hormones subsequently to appropriate clean up in cattle urine.

The two isomers (*cis* and *trans*) of the mycotoxin zearalenone can be determined down to 5 ng/g in 10 g cereal samples. This is of particular importance, be-

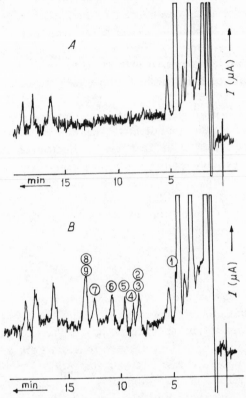

Fig. 20. — Trace determination of substances with oestrogenic effects in chicken meat. Column Zorbax CN5, glassy carbon electrode in Metrohm cell for voltam-metric HPLC-detection at +1.2 V (Ag AgCl). Polarograph Princeton Applied Research Corp. PAR 364, sensitivity 1 μA full scale. (A) chicken meat blank; (B) chicken meat containing about 800 pg anabolic drug in 1 g meat sample. (1) estriol, (2) estradiol-17β, (3) zeranol, (4) estradio-17α, (5) estrone, (6) zearale-none, (7) Dienestrol, (8) Diethylstilbestrol, (8) Hexestrol

cause the *trans*-isomer has probably the higher toxicity, due to its structural similar-ities to estrogenic substances, in contrast to *cis*-zearalenone (see Fig. 21) [61].

5.2. *Investigations and assessment of toxic heavy metals*

The borderlines between problems belonging into the disciplines of biology and biochemistry or ecological chemistry are necessarily and obviously floating to a certain extent. Therefore, in a number of cases aspects from both disciplines

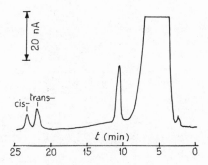

Fig. 21. — HPLC-VD of ground corn sample incubated for 3 days with 1 μg of transzearalenone. Lichrosorb RP-8 column, 45:55 acetonitrile/water, 0.05 M $LiClO_4$, flow rate 1.2 cm^3 min^{-1}, glassy carbon electrode +1.2 V (Ag AgCl)

are intercorrelated. Among the considerable number of various environmental chemicals with biological impact certain heavy metals, *e.g.* Cd, Pb, Hg, Ni, As, Cu and others have gained particular significance. At present a task of high priority and significance is to determine and monitor the content of these heavy metals in natural waters, soil, food and all kinds of biological materials in order to assess how far eventually their levels exceed as consequence of anthropogenic emissions natural baselines or with respect to toxicological aspects tolerable thresholds. As these metals have a considerable toxicity and tend to be accumulated by living systems also rather small concentrations require careful monitoring. Suitable voltammetric methods play an increasingly eminent part in this demanding and challenging branch of trace chemistry and ultra trace analysis [62]. In all types of natural waters differential pulse stripping voltammetry and differential pulse voltammetry have become the most superior approach with respect to sensitivity and reliability for the determination of the above mentioned toxic heavy metals. A flow chart of the analytical procedure is shown in Fig. 22.

Moreover, due to the fact that voltammetry is a species-sensitive rather than only an element-specific method it offers very versatile opportunities to elucidate the distribution of the overall metal concentration in a natural water type over the various kinds of species formed with inorganic and organic ligands [63, 64]. Metal quantities and their speciation pattern have key significance for their uptake by acquatic organisms.

Also for the reliable assessment of toxic heavy metals in soil, plants, organisms, tissues, body fluids and all types of food, voltammetry in the differential pulse mode provides in connection with suitable and rather uniformly applicable digestion procedures one of the most appropriate and certainly comparatively most

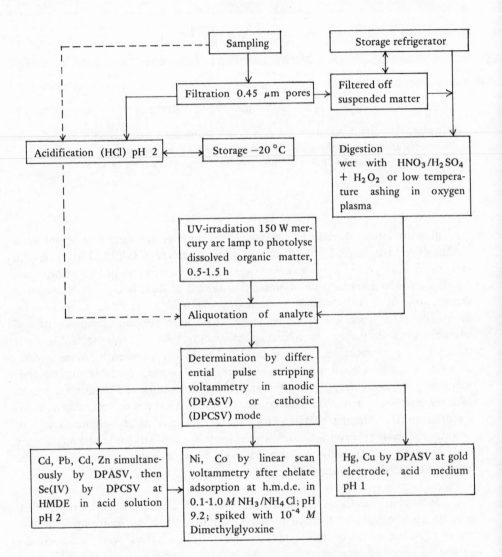

Fig. 22. — Flow chart of analytical procedure for toxic heavy metal analysis in natural waters (rivers, lakes, rain, sea and drinking water) and waste waters. For certain water types (samples from most parts of the open oceans where biological productivity is low; rain water if not heavily polluted by organics and usually drinking water) certain stages can be by-passed as is indicated by the dotted lines

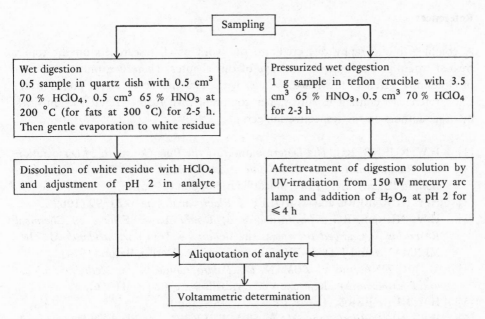

Fig. 23. — Flow chart of analytical procedure for toxic heavy metal analysis in biological materials with different digestion alternatives

reliable alternative [65, 66]. The flow chart in Fig. 23 reflects the overall analytical procedure for biological matrices. Once the stage of the aqueous analyte has been reached the subsequent voltammetric determination of the toxic trace metals proceeds similarily to that in natural water samples (Fig. 22).

6. Concluding remarks

Although only a selection of different examples could be treated in this chapter, it can be concluded in general that advanced voltammetric methods provide indispensable versatile and significant potentialities to extend and deepen the understanding of the bioelectrochemical and biophysicochemical aspects of the behavior and fate of substances and chemicals of biological significance in biochemical processes going on in living systems. A new dimension of potentialities stems from the concerted application of voltammetric and spectroscopic methods.

References

A complete bibliography due credit to the work of all researchers on the topics treated would go far beyond the scope of this chapter. Therefore, only a selection of citations is given to provide the reader with reference access to the points made. Cited references contain usually rather comprehensive further references on work by other authors on the respective subject.

[1]   H.W. NÜRNBERG, *The Determination of the Rate Constants of Dissociation and Recombination for Carboxylic Acids by High Level Faradaic Rectification*, in *Polarography 1964*, MacMillian, London (1966).

[2]   H.W. NÜRNBERG and G. WOLFF, *J. Electroanal. Chem.* **21**, 99 (1969).

[3]   H.W. NÜRNBERG, *The Influence of Double-Layer Effects on Chemical Reactions at Charged Interfaces*, in *Membrane Transport in Plants*, U. ZIMMERMANN and J. DAINTY (Editors), Springer Verlag, Berlin (1974).

[4]   M. BREZINA and P. ZUMAN, *Die Polarographie in der Medizin, Biologie und Pharmazie*, Akademische Verlagsgesellschaft, Leipzig (1956).

[5]   H.W. NÜRNBERG, *Angew. Chem.* **72**, 433 (1960).

[6]   H.W. NÜRNBERG and M. v. STACKELBERG, *J. Electroanal. Chem.* **2**, 181 (1961), **4**, 1 (1962).

[7]   H.W. NÜRNBERG, *Fresenius Z. Anal. Chem.* **186**, 1 (1962).

[8]   P. ZUMAN, *Organic Polarographic Analysis*, Pergamon Press, Oxford (1964).

[9]   J. HEYROVSKY and J. KUTA, *Grundlagen der Polarographie*, Akademie Verlag, Berlin (1965).

[10] P. ZUMAN, *Substituent Effects in Organic Polarography*, Plenum Press, New York (1967).

[11] H. HOFFMANN, *Polarographic Analysis in Pharmacy*, in *Electroanalytical Chemistry*, H.W. NÜRNBERG (Editor), J. Wiley, New York (1974).

[12] H. JEHRING, *Elektrosorptionsanalyse mit der Wechselstrompolarographie*, Akademie Verlag, Berlin (1974).

[13] H.W. NÜRNBERG and B. KASTENING, *Polarographic and voltammetric techniques*, in *Methodicum Chimicum*, F. KORTE (Editor), Academic Press, New York (1974), Vol. 1A.

[14] L. MEITES, H.W. NÜRNBERG and P. ZUMAN, *Pure Appl. Chem.* **45**, 81 (1976).

[15] W.F. SMYTH (Editor), *Polarography of Molecules of Biological Significance*, Academic Press, New York (1979).

[16] W.F. SMYTH (Editor), *Electroanalysis in Hygiene, Environmental, Clinical and Pharmaceutical Chemistry*, Elsevier, Amsterdam (1980).

[17] A.M. BOND, *Modern Polarographic Methods in Analytical Chemistry*, M. Dekker, New York (1980).

[18] H.W. NÜRNBERG, *Differentielle Pulspolarographie, Pulsvoltammetrie und Pulsinversvoltammetrie*, in *Analytiker-Taschenbuch*, R. BOCK, W. FRESE-NIUS, H. GÜNZLER, W. HUBER and G. TÖLG (Editors), Springer Verlag, Berlin (1981), Vol. 2.

[19] B. JANIK and P. ELVING, *Chem. Rev.* 68, 295 (1968).

[20] D. KRZNARIC, P. VALENTA and H.W. NÜRNBERG, *J. Electroanal. Chem.* 65, 863 (1975).

[21] Y.M. TEMERK, P. VALENTA and H.W. NÜRNBERG, *Bioelectrochem. Bioenerg.* 7, 705 (1980).

[22] K.M. ERVIN, E. KOGLIN, P. VALENTA and H.W. NÜRNBERG, *J. Electroanal. Chem.* 114, 179 (1980).

[23] P. VALENTA, H.W. NÜRNBERG and D. KRZNARIC, *Bioelectrochem. Bioenerg.* 3, 418 (1976).

[24] Y.M. TEMERK, P. VALENTA and H.W. NÜRNBERG, *J. Electroanal. Chem.* 100, 77 (1979).

[25] Y.M. TEMERK, P. VALENTA and H.W. NÜRNBERG, *J. Electroanal. Chem.* 100, 289 (1980).

[26] D. KRZNARIC, P. VALENTA, H.W. NÜRNBERG and M. BRANICA, *J. Electroanal. Chem.* 93, 41 (1978).

[27] P. VALENTA and D. KRZNARIC, *J. Electroanal. Chem.* 75, 437 (1977).

[28] Y.M. TEMERK, P. VALENTA and H.W. NÜRNBERG, *J. Electroanal. Chem.* 131, 265 (1982).

[29] P. VALENTA, H.W. NÜRNBERG and P. KLAHRE, *Bioelectrochem. Bioenerg.* 2, 204 (1975).

[30] H.W. NÜRNBERG and P. VALENTA, *Croat. Chem. Acta* 48, 623 (1976).

[31] H.W. NÜRNBERG and P. VALENTA, *Bioelectrochemical Behavior and Deconformation of Native DNA at Charged Interfaces*, in *Ions in Macromolecular and Biological Systems*, D.H. EVERETT and B. VINCENT (Editors), *Proc. 29th Colston Symp.*, Scientechnica, Bristol (1978).

[32] J.M. SEQUARIS, E. KOGLIN, P. VALENTA and H.W. NÜRNBERG, *Ber. Bunsenges. Phys. Chem.* 85, 512 (1981).

[33] H.W. NÜRNBERG, *Untersuchungen zur Katalyse der Wasserstoffabscheidung durch Organische Stickstoffbasen an der Quecksilberkathode*, in *Advances in Polarography*, I.S. LONGMUIR (Editor), Pergamon Press, Oxford (1960), Vol. 2.

[34] P. VALENTA and P. GRAHMANN, *J. Electroanal. Chem.* 49, 41 (1974).

[35] P. VALENTA and H.W. NÜRNBERG, *Biophys. Struct. Mechanism* 1, 17 (1974).

[36] B. MALFOY, J.M. SEQUARIS, P. VALENTA and H.W. NÜRNBERG, *Bioelectrochem. Bioenerg.* 3, 440 (1976).

[37] E. PALECEK, *Collect. Czechoslov. Chem. Commun.* 39, 3449 (1974).

[38] E. PALECEK, *Bioelectrochem. Bioenerg.* **8**, 469 (1981).

[39] J.M. SEQUARIS, P. VALENTA, H.W. NÜRNBERG and B. MALFOY, *On the Interfacial Behavior of Double Stranded Polynucleotides in Alkaline Solution,* in *The Behavior of ions on Macromolecular and Biological Systems,* D.H. EVERETT and B. VINCENT (Editors), *Proc. 29th Colston Sym.* Scientechnica, Bristol (1978).

[40] B. MALFOY, J.M. SEQUARIS, P. VALENTA and H.W. NÜRNBERG, *J. Electroanal. Chem.* **75**, 455 (1977).

[41] T.L. HILL, *J. Am. Chem. Soc.* **80**, 2142 (1958).

[42] J.M. SEQUARIS, B. MALFOY, P. VALENTA and H.W. NÜRNBERG, *Bioelectrochem. Bioenerg.* **3**, 461 (1976).

[43] E. KOGLIN, J.M. SEQUARIS and P. VALENTA, *J. Mol. Struct.* **60**, 421 (1980).

[44] E. KOGLIN, J.M. SEQUARIS and P. VALENTA, *Z. Naturforsch.* **36**, 809 (1981).

[45] J.M. SEQUARIS, P. VALENTA, H.W. NÜRNBERG and B. MALFOY, *Bioelectrochem. Bioenerg.* **5**, 483 (1978).

[46] J.M. SEQUARIS, P. VALENTA and H.W. NÜRNBERG, *Intl. J. Radiat. Biol. Relat. Stud. Phys. Chem. Med.*, in press.

[47] J.M. SEQUARIS, *Applications de la voltammétrie à l'étude des acides désoxyribonucléiques natifs et modifiés,* Thèse Doctorat d'Etat, Univ. Orléans (1982).

[48] J.M. SEQUARIS, P. VALENTA and H.W. NÜRNBERG, *J. Electroanal. Chem.* **122**, 263 (1981).

[49] V. BRABEC and J. KOUDELKA, *Bioelectrochem. Bioenerg.* **7**, 793 (1980).

[50] L. TRNKOVA, M. STUDNIKOVA and E. PALECEK, *Bioelectrochem. Bioenerg.* **7**, 643 (1980).

[51] V. BRABEC, *Bioelectrochem. Bioenerg.* **8**, 437 (1981).

[52] M.Z. WRONA, R.N. GOYAL and G. DRYHURST, *Bioelectrochem. Bioenerg.* **7**, 433 (1980).

[53] A. BRAJTER-TOTH, R.N. GOYAL, M.Z. WRONA, T. LACAVA, N.T. NGUYEN and G. DRYHURST, *Bioelectrochem. Bioenerg.* **8**, 413 (1981).

[54] J.A. REYNAUD, P.I. SICARD and A. OBRENOVITCH, *Experienta Suppl.* **18**, 543 (1971).

[55] E. LUKASOVA, M. VOJTISKOVA and E. PALECEK, *Bioelectrochem. Bioenerg.* **7**, 671 (1980).

[56] C. MOLINIER-JUMEL, B. MALFOY, J.A. REYNAUD and G. AUBEL-SADRON, *Biochem. Biophys. Res. Commun.* **84**, 441 (1978).

[57] H. BERG, G. HORN and U. LUTHARDT, *Bioelectrochem. Bioenerg.* **8**, 537 (1981).

[58] M.R. SMYTH, C.G.B. FRISCHKORN and H.W. NÜRNBERG, *Anal. Proc.* **18**, 215 (1981).

[59] C.G.B. FRISCHKORN, M.R. SMYTH, H.E. FRISCHKORN and J. GOLI-MOWSKI, *Fresenius Z. Anal. Chem.* **300**, 407 (1980).

[60] M.R. SMYTH and C.G.B. FRISCHKORN, *Fresenius Z. Anal. Chem.* **301**, 220 (1980).

[61] M.R. SMYTH and C.G.B. FRISCHKORN, *Anal. Chim. Acta* **115**, 292 (1980).

[62] H.W. NÜRNBERG, *Pure Appl. Chem.* **54**, 853 (1982).

[63] H.W. NÜRNBERG and B. RASPOR, *Environ. Technol. Letters* **2**, 457 (1981).

[64] H.W. NÜRNBERG and P. VALENTA, *Potentialities and Applications of Voltammetry in Chemical Speciation of Trace Metals in the Sea*, in *Trace Metals in Sea Water*, C.S. WONG and K. BRULAND (Editors), Plenum Press, New York, in press.

[65] H.W. NÜRNBERG, *A Critical Assessment of the Voltammetric Approach for the Study of Toxic Metals in Biological Specimens and Their Ecosystems*, in *Electroanalysis in Hygiene, Environmental, Clinical and Pharmaceutical Chemistry*, W.F. SMYTH (Editor), Elsevier, Amsterdam (1980).

[66] H.W. NÜRNBERG, *Potentialities and Applications of Voltammetry in the Analysis of Toxic Trace Metals in Body Fluids*, in *Analytical Techniques for Heavy Metals in Body Fluids*, S. FACCHETTI (Editor), Elsevier, Amsterdam, in press.

[67] S. LEWIN, *Displacement of Water and Its Control of Biochemical Reactions*, Academic Press, New York (1974).

[68] R.N. ADAMS, *Anal. Chem.* **48**, 1126 A (1976).

[69] R.F. LANE, A.T. HUBBARD, C.D. BLAHA, *Bioelectrochem. Bioenerg.* **5**, 504 (1978).

[70] H.Y. CHENG, W. WHITE and R.N. ADAMS, *Anal. Chem.* **52**, 2445 (1980).

TRANSMEMBRANE POTENTIAL AND REDOX REACTIONS FROM A PHYSIOLOGICAL POINT OF VIEW*

MARTIN BLANK

*Department of Physiology, College of Physicians and Surgeons
Columbia University, New York, N.Y. 10032, U.S.A.*

Contents

* Supported by research grant PCM 78-09214 from the NSF and N0014-80-C-0027 from the ONR.

1. Introduction

The many sub-disciplines of biology frequently overlap, but each is usually characterized by its special approach to the problems of living systems. The unique point of view of physiology is the emphasis on biological *function*, and the goal of physiologists is to explain processes (*e.g.*, transport across membranes and excitation) in terms of the laws of physics and chemistry. The information that is obtained by the other disciplines about morphology or chemical composition are important components of an explanation, but in the end the physiologist will use this information to shed light on a process involved in function.

These lectures deal with the group of problems relating to the role of membrane potentials in physiological processes. The composition and structure of the membrane will be considered as well as information about the laws of physical chemistry that apply to membrane processes. Finally, these ideas will be brought together to consider how membrane potentials arise and what factors affect their magnitudes. The presentation will be general, and the references given will tend to be of reviews rather than original papers.

2. Natural membranes

Natural membranes differ from one cell to another, and even in a single cell, the plasma membrane differs from those of the organelles. If in addition, one considers the regular turnover of membrane components and the changes with cell age, it is obvious that many different structures are covered by the same term. Nevertheless, we speak of a generalized natural membrane, that is about 8 nm (80 Å) thick, because there appears to be a basic structure that is present in all cases.

Much appears to be known about the composition, structure and physical properties of the erythrocyte membrane, and many use this system as the prototype of a natural membrane. Most investigators agree that the membrane consists of an asymmetric lipid bilayer (choline phospholipids and glycolipids on the external face and amino phospholipids on the cytoplasmic face), and an

asymmetric distribution of membrane protein, with the major portion on the
cytoplasmic side and some going through the bilayer. There are approximately
equal amounts of lipid and proteins, and the proteins appear to form a network that provides structural stability [1].

It should be emphasized that although there is general agreement about
the foregoing statement, there are very different pictures in the minds of biologists when they use the term *membrane*. Table 1 indicates the wide diversity
in the points of view of the different experimental approaches to membrane
properties. The multiplicity of approaches is useful because it leads to complementary types of information. Furthermore, since these are views of the same
membrane, there must be no contradictions. Therefore, one must be able to
account for a physical property of the natural membrane in terms of its molecular components and their arrangement. Many biological scientists go so far
as to say that all one really needs to know is the *structure* of a system, and an
understanding of the function will follow.

Before leaving this topic, I would like to discuss the problem of how to
classify the natural membrane as a state of matter. It is too small to be considered a bulk system, but too thick to be called a film [although the 8 nm
(80 Å) thick structure contains several layers composed of lipids and proteins,
it apparently functions as a unit]. Recent studies of membrane protein films
of increasing thickness suggest that the membrane structure is probably closer
to a bulk phase, and that the actual membrane dimension optimizes a number
of physical properties with opposing dependences on thickness [2].

Summarizing the results of measurements on membrane protein films
of different thicknesses, it appears that there are two important transition
points (see Fig. 1 and 2) in the variation of physical properties:

1)   At a single monolayer, of about 2 nm thickness, the film develops a
      yield, *i.e.* it is able to sustain mechanical stresses. There is also a very
      low permeability to ions due to close packing of the charged groups
      at the interface. As soon as one makes the film thicker the additional
      molecules disrupt the structure, increase the permeability and apparently
      change the permeability mechanism. The transition at one monolayer
      also appears (approximately) in the surface potential and in the surface
      energy. (The surface free energies of dissolved biopolymers and membrane
      proteins can be calculated in terms of the surface layer only).

2)   At 5-8 nm about 3-4 monolayers, five surface properties show tran-
      sitions. The two-dimensional yield is a maximum, while the surface shear
      resistance, surface potential, surface viscosity and elastic modulus reach
      plateaus in this range.

Table 1. — Different views of the natural membrane

Approach	Membrane composition	Arrangement of components	Function
Biochemistry	Polar lipids and polymers (*e.g.*, proteins)	Chemically asymmetric bilayer. Specialized molecules for structure, transport	Matrix for enzyme reactions, inter-molecular inter-actions
Morphology	Components that react with heavy metals and oil soluble reagents	Asymmetric ultrathin layer, split to reveal internal structures	Boundary of cell or organelle
Physiology (transport)	Semipermeable array with special molecules (*e.g.*, pump enzyme, carrier)	Permeability barrier with selective structures (*e.g.*, enzyme, pore, carrier)	Separation and concentration of cell constituents
Physiology (electrophysiology)	Resistance, capac-itance, and ion channels with gates)	Equivalent circuit of electrical elements	Electrical prop-erties (*e.g.*, excitation)
Electrochemistry	Dielectric materials, charges (ions and dipoles)	Phase boundaries, asymmetric charge distribution	Charge transport and variations of membrane potential
Rheology	Elastic, viscous, and structural elements	Equivalent circuit of rheological elements	Strain, yield and flow under mechanical stresses

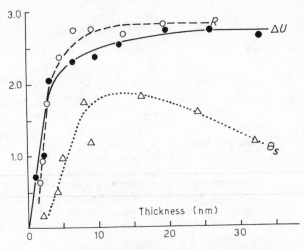

Fig. 1. — Three surface properties of red cell membrane protein layers as a function of the thickness of the film in nm (1 nm = 10Å). The surface potential ΔV, in millivolts (•); the surface yield, θs, in dynes per centimeter ($\triangle$); and the shear resistance, R, in arbitrary units (○). All properties are at 25°C and on 0.1M NaCl. (Ordinate • = 10^{-2} mV; $\triangle$ = 10^{-2} Dynes/cm; ○ = 10^{-1} arbitrary units). (Reproduced with permission from the *J. Colloid Interface Sci*).

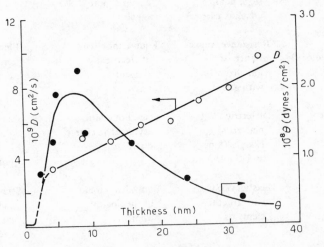

Fig. 2. — Two bulk properties of red cell membrane protein multi-layers as a function of the thickness of the film in nm (1 nm = 10Å). The diffusivity, D, in square centimeters per second (○) and the yield, in dynes per square centimeter (•). Below 2 nm the straight line indicates the range of D values obtained by another technique, and the dashed line joins the two sets of measurements of the diffusivity of the films to ions. (Reproduced with permission from the *J. Colloid Interface Sci.*).

From this, it appears that the membrane thickness is in a range where many physical properties have stabilized. Further increases in thickness will not cause any qualitative changes, but only result in a greater permeability barrier and a proportionally weaker structure. This suggests that the membrane thickness may represent an optimum. The membrane is at a point where we have reached stable properties, optimal mechanical strength and a diffusion barrier that is not so great as to preclude the exchange of essential substances with the environment. In addition, the presence of proteins at the surfaces of a membrane enables them to engage in specific reactions whose energetics are governed by the surface layer only.

3. Electrical double-layers

The lipids and proteins in natural membranes contain many groups (*e.g.* phosphate, carboxylate) that are charged under normal body conditions, and the charge has been measured on some membranes. For example, it has been shown that there is a net negative charge on the surface of a red cell as well as on the inner and outer faces of a nerve axon membrane. Since these membranes are normally in contact with aqueous solutions containing dissolved salts, the charged groups on the surface must interact with counter-ions in the solution to form electrical double-layers.

The theory of electrical double-layers has been used for years and accounts for many of the properties of charged surfaces. The original ideas were based on many approximations (*e.g.* point charges, uniform surface charge density, planar surface) which have now been corrected, but the basic ideas of the elementary theory still lead to reasonable qualitative conclusions and are useful for understanding effects at surfaces.

Without going into detail we can point out some of the conclusions of electrical double-layer theory that are important for understanding membrane properties:
1) The concentrations of adsorbed counterions is increased and that of co-ions decreased at the charged surface. The concentration at the surface relative to that in the bulk is actually determined by an exponential (BOLTZMANN) factor that is a function of the electrical potential at the surface.
2) The surface has the properties of an electrical circuit element: the capacitor. It can store charge because when one charges up the surface, counterions adsorb. In fact, the capacitance of these surfaces is generally very large.

3)   The thickness of the electrical double-layer region depends on the physical properties of the solution, but it could be rather small. In physiological saline the double-layer is less than 1 nm thick. (This is the distance where the potential falls to 1/e of the value at the charged surface). These small dimensions indicate that important changes occur over very small distance.

4.   Membrane processes: special effects in surface regions

4.1.   Transport through ultrathin layers

Because the thickness of a natural membrane is comparable to molecular dimensions, its permeability is essentially different from that of a macroscopic phase. In general, physical laws are based on averages, and they are valid because there are many particles and many events involved. In membrane permeation the size of the permeant is comparable to the thickness of the barrier, and one cannot assume a multi-collision process with the same averaging as in macroscopic diffusion.

Insight into the important factors in membrane permeability can be obtained from studies of lipid monolayers, which are comparable to natural membranes in size and structure. In these studies the permeability varies directly with the amount of free space. The least permeable monolayers are of the close-packed incompressible type, and the most freely permeable form relatively open structures such as those of protein, cholesterol or oleic acid films. In a homologous series, the permeability varies inversely with the length of the hydrocarbon chain, and in a group of compounds having the same chain length, the permeability depends on the size of the polar group relative to that of the non-polar chain [3].

Monolayer permeation resembles diffusion in macroscopic solid phases, since the permeability depends upon the size and number of holes available in a lattice. There is also interference between a permeant and other gaseous components of the system that resemble diffusion in a fixed lattice. (For example, permeability to carbon dioxide is affected by water vapor that is present in the monolayer). The variation of the activation energy with monolayer thickness implies a one-step discontinuous permeation as in solid diffusion from hole to hole. The discontinuous nature is in line with the absence of interfacial partition equilibrium effects (i.e. there is no apparent dissolution of the permeant within the monolayer) during permeation (see Fig. 3).

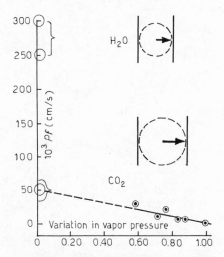

Fig. 3. — The permeabilities of an octadecanol monolayer to H$_2$O and to CO$_2$ are indicated on the ordinate, and the approximate cross-sectional areas of the two molecules are also shown. The permeability of an octadecanol monolayer to CO$_2$ varies with the vapor pressure of the aqueous phase upon which it is spread, indicating interference between the two gases, and these data are plotted on the lower part of the figure. The permeabilities are determined by the sizes of the gases and not by their solubilities. (Reproduced with permission from the *J. Gen. Physiol.*).

It therefore appears that a lipid monolayer can be thought of as a solid phase for the purpose of understanding the permeation process, but one in which the number of events involved in a one-step jump across the layer is not sufficient to allow for averaging. If monolayer permeation is formulated in terms of spaces that result from the natural free area due to the packing as well as from the equilibrium fluctuations in monolayer density, the equations account for the variation of the permeation rate with temperature and permeant size and also predict the self diffusion coefficient and the frequency of phospholipid flip-flop in a bilayer membrane [4].

4.2. Ion transport and surface charge

The surface charge is the main factor that affects ion transport across thin films, and the relation between the charge density and the resistance to ion transport has been demonstrated in experimental systems. In the transport

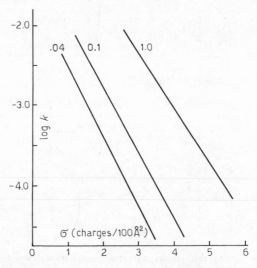

Fig. 4. — The logarithm of the ion transport rate constant, k (in cm/s) of Cu^{2+} ion diffusing through a decylammonium monolayer at a mercury water interface, plotted against σ, the surface charge density (in charges /100 Å^2). The numbers on the lines indicate the ionic strength of the aqueous phase containing 10^{-3} M Cu^{2+} ions. (Reproduced with permission from *Bioelectrochem. Bioenerg.*).

of cations through oriented charged monolayers at a mercury water interface, the logarithm of the rate constant for passing through the monolayer was found to be linearly related to the surface charge (Fig. 4). Also, the variation of the resistance of a positively charged cholesterol bilayer with the concentration of long chain cations in the adjacent aqueous phases appears to be similar to the monolayer results. The experimental relation between the logarithm of the conductance and the surface charge can be derived from electrical double layer theory [5]. It appears that ion flow depends upon the concentration of ions in the surface layer, and the surface charge exerts its influence by controlling the concentration of ions in the double layer. The control of ion flow is therefore a partition mechanism involving equilibration of the surface layer with the bulk solution. We shall explore the implications of these and related observations for a membrane system, when we discuss the Surface Compartment Model.

4.3. Membrane reactions: ion and drug binding

The interactions of substances with specific components of natural membranes have been studied from the point of view of membrane *receptors* in drug reactions, membrane *carriers* in transport, and membrane *enzymes* in biochemical reactions. All of these studies involve the interaction of a dissolved chemical with a specialized component of a membrane. It is therefore not surprising that all the different types of membrane reactions show saturation interaction curves (indicating a limited number of membrane components) and competition between substances having related structures (indicating structural specificity). The idea of saturation, coupled with mass action, has provided the essential theoretical framework for analyzing membrane reactions.

Another characteristic of membrane reactions is the frequent occurrence of cooperative effects, *i.e.* interactions where there are changes in the affinity as a reaction proceeds. It is easy to understand saturation and competitions in terms of a limited number of membrane sites with rather specific structures. Cooperative interactions can be understood in terms of the lack of independence of sites on a surface, and the additional free energy changes that result from changes in membrane structure. For example, the interaction of charged substances with oppositely charged groups on a membrane always leads to a decrease in the free energy and consequently to an increase in the affinity. An uncharged substance can also affect the free energy of the system if it interferes with the membrane structure. If one analyzes the factors that affect the magnitude of the free energy, it appears that the affinity of a membrane reaction would vary with the surface charge and the ionic strength of the medium [6].

The surface of the membrane provides the link between the membrane reaction and the observed physiological properties. To a membrane physiologist, the stimulation of a cell usually leads to excitation or secretion, processes that are generally initiated by a change in the membrane potential, which in turn is due to an altered permeability. Our studies on model systems indicate that ion binding accompanied by a change in surface charge can lead to a change in ion permeability. Any membrane reaction should alter the permeability, but changes in charge are especially significant since they influence ion concentrations in the surface and ion transport.

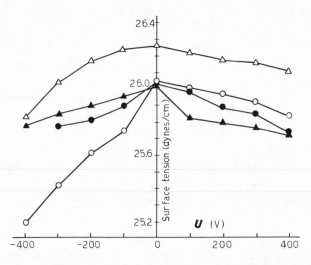

Fig. 5. — The interfacial tension as a function of the electric field imposed across the water nitrobenzene interface when 10^{-3} M amino acid is dissolved in the water at various pH's. The open symbols are for glycine (O) and
for glutamic acid (△) at pH's below their isoelectric points. The shaded symbols are of the same amino acids above their isoelectric points. (Reproduced
with permission from *Science*).

4.4.   *Concentration changes during current flow*

When an electric field is imposed across a liquid-liquid interface one can
expect a change in the interfacial concentration of ions by either of two
mechanisms. The better known mechanism is electro-capillary adsorption, where
the electric field causes an adsorption of cations or anions depending on the
charge of the surface. In this case there is negligible transference through the
interface. The second mechanism that gives rise to changes (increases as well
as decreases) in interfacial concentration is operative when there is transference
through the interface. The change in concentration occurs at the interface but,
owing to diffusion, extends into both phases as time passes [7].

Both types of adsorption can be demonstrated in the water-nitrobenzene
systems, two immiscible fairly high dielectric materials in which one can dissolve a variety of ionic solutes. The two mechanisms depend upon solubility
relations and lead to changes in the interfacial concentration of ions. If the ion
is soluble in the aqueous phase only, it behaves as in electrocapillary adsorption
(see Fig. 5). If it is soluble in both phases, it exhibits increases and decreases

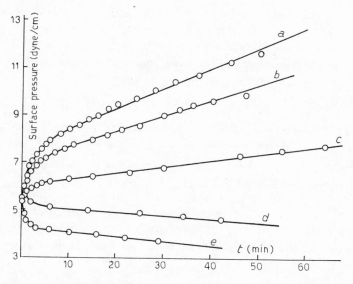

Fig. 6. – The variation in surface pressure with time at a water-nitrobenzene interface when different currents are passed through the interface. The
values of the current in microamperes are: a, 150; b, 106; c, 52; d, −93; and
e, −208, and the long chain cationic surfactant, CTAB, is initially 0.96 x 10^{-4}
M in water and 1.55 x 10^{-4} M in nitrobenzene, the equilibrium partition at 25
°C. (Reproduced with permission from *Science*).

in concentration, depending upon the ion transport numbers and the direction
of the current (see Fig. 6). Electrocapillary adsorption is obviously the limiting case of the more general mechanism.

5.   Association of multi-subunit proteins

In the previous section, it was shown that many membrane processes are
directly affected by changes in the charge and in the ionic concentrations at
the surfaces. Indirect effects are also possible because many membrane proteins are multi-subunit aggregates that can exist in different degrees of association with different properties (*e.g.* permeability to ions, mechanical strength).
Several red cell membrane proteins, including spectrin, are associated in a
structural network on the cytoplasmic face [1]. Of the membrane components
directly associated with transport processes, the Na-K ATPase of the kidney is
composed of four subunits, each of which spans the membrane. The latest
studies on the acetylcholine receptor indicate that it has five subunits, two of
which are identical and all of which have contact with the aqueous phase at

the outer face of the membrane. While it is difficult to describe the inter-actions between the protein subunits and the lipids at the interior of the mem-brane, the interactions in the aqueous phase are similar to those between subunits of globular proteins in solution. In all of these cases it is expected that the mechanical strength and the permeability depend on the degree of association of subunits.

Although oligomeric proteins exist in many sizes, shapes and degrees of aggregation as a result of specific interactions, there appear to be some com-mon responses of the different systems to changes in physical conditions. Thus, *aggregation* generally increases as one increases the monomer concentra-tion, and *disaggregation* increases as the pH is further away from the isoelectric point or the ionic strength increases. These general observations are consistent with accepted theoretical ideas of mass action and electrostatic repulsion (al-though the ionic strength effect appears to be the opposite of what is expected), and suggest that non-specific factors play an important role in these equilibria.

On looking at molecular models of proteins one is struck by their simi-larity to classic micelles or emulsion droplets, *i.e.* the hydrophobic residues are on the inside and the polar residues on the surface. Non-protein structures are frequently stabilized by surface films causing low interfacial free energies, and the point of instability can be determined from the changes in interfacial free energy. Since proteins behave like hydrophobic materials stabilized by a hydro-philic surface film, it should be possible to estimate changes in the interfacial free energy due to changes in pH or ionic strength and thereby estimate the point of aggregation. This in effect was done in the case of hemoglobin. We calculated changes in the interfacial free energy on the basis of changes of molecular area (estimated from X-ray data) and charge (estimated from titra-tion data), and determined the pH's at which the tetramer will dissociate into two dimers. This work is illustrated in Fig. 7 and 8, and we can see that the agreement is quite good [8].

Hemoglobin is an easy system to consider because it is relatively simple and much information is available about its properties. However, it appears that more complex systems, both in terms of size (*e.g.* hemocyanin) and heter-ogeneity of interactions (*e.g.* membrane proteins), have many of the same properties. For example, the properties of hemoglobin with a molecular weight of 67 000 and 4 subunits, are similar to those of the hemocyanins, which have a molecular weight of 3 300 000 to 8 000 000 daltons and many more subunits. Hemoglobins and hemocyanins both dissociate at acid and alkaline pH's and at high ionic strength, and also show similar differential titration curves between the oxy and deoxy forms of the molecule.

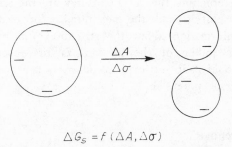

$$\Delta G_S = f\,(\Delta A, \Delta \sigma)$$

Fig. 7. — The model used for calculating the free energy of the dissociation of a hemoglobin tetramer into two dimers. The diagram illustrates the changes in interfacial area, ΔA, and surface charge density, $\Delta \sigma$, that result from the reaction. (Reproduced with permission from *Coll. Surf.*).

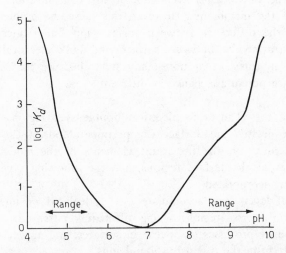

Fig. 8. — The variation of the dissociation constant, K_d (on a logarithmic scale), with pH, assuming $K_d = 1$ for pH = 6.8, the isoelectric point of de-oxygenated hemoglobin. The ranges indicated below the curve in both the acid and alkaline regions correspond to the calculated range of dissociation into dimers, which is close to observations. The change in the magnitude of K_d is also close to observations. (Reproduced with permission from *Coll. Surf.*).

From our results and related work it appears that the energetics governing these dissociation reactions can be understood in terms of the surface free energy that can be calculated for each oligomeric structure. The two most important factors determining the surface energy are the area of contact of the molecule with the solution and the net number of charged groups on the surface. Increased information about the structure of membrane proteins coupled with an understanding of the physical factors governing association of protein subunits should help to explain observed changes in membrane permeability and membrane potential.

6.   Membrane potentials

Most physiologists accept the idea that membrane potentials are non-equilibrium electrical potentials that are generated by the chemical potential differences of the permeable ions across the membrane. The steady state that results from the balance between chemical and electrical forces enables the calculation of the membrane potential in terms of the concentrations and permeabilities (transport numbers) of the ions. The ionic concentrations that are used are the bulk values even though the concentrations are different at the surfaces of the membranes. However, this causes no problems in the steady state, because the difference between surface and bulk concentrations is balanced by the difference between surface and bulk electrical potentials. The problems arise in transient or non-steady states because the surface concentrations and surface potentials change at different rates.

To study the changes in electrical double-layer regions during transient states, we have devised the Surface Compartment Model (SCM) for the steady state [9]. Our ideas on the important elements of the model developed as a result of studies of the surface regions, and the following experimentally based ideas have been incorporated:
1)   Interfacial layers can accumulate or be depleted of ions as a result of current flow across regions having different ion transport numbers.
2)   Ion exchange processes can occur in surfaces, and appreciable ion currents can result from the originally bound ions.
3)   Ion transport across monolayers and bilayers is strongly dependent upon the surface charge density and the ionic strength of the solutions.

We have adapted the SCM equations for dealing with transient states (BLANK and KAVANAUGH [10]), such as voltage clamp conditions, where in addition to ion flow there is cation surface binding and membrane charge transfer (*i.e.*, gating currents). The result is a highly coupled system of 14 independent differential equations. Six differential equations describe the time variation of the concentrations of the three ionic species in the two surface compartments, four equations are for the cation binding at the two surfaces, two equations describe changes in surface charge, and two equations are for the surface potentials. These equations are given below in terms of a membrane system that consists of five discrete regions:

    — an outside bulk (reservoir) phase, compartment 1,
    — an outer surface compartment 2 with capacitance $C1$,
    — a membrane having surface charges $Q2$ and $Q3$ on the two faces, and a dielectric capacitance $C2$,
    — an inner surface compartment 3 with capacitance $C3$,
    — an inner bulk (reservoir) phase, compartment 4.

Changes in the ion concentrations $[Na^+]$, $[K^+]$ and $[A^-]$ are given by:

$$[\dot{Na}^+]2 = (1/L2)*(JNa^+1 - JNa^+ - PNa^+ - [\dot{Na}^+]22) \tag{1}$$

$$[\dot{K}^+]2 = (1/L2)*(JK^+1 - JK^+ - PK^+ - [\dot{K}^+]22) \tag{2}$$

$$[\dot{A}^-]2 = (1/L2)*(JA^-1 - JA^- - PA^-) \tag{3}$$

$$[\dot{Na}^+]3 = (1/L3)*(JNa^+ + PNa^+ - JNa^+3 - [\dot{Na}^+]33) \tag{4}$$

$$[\dot{K}^+]3 = (1/L3)*(JK^+ + PK^+ - JK^+3 - [\dot{K}^+]33) \tag{5}$$

$$[\dot{A}^-]3 = (1/L3)*(JA^- + PA^- - JA^-3) \tag{6}$$

where the points over the symbols of the concentration stand for the derivatives $d[Na^+]/dt$ etc.

In the equations, L's are compartment thicknesses, J's are fluxes given by the NERNST-PLANCK equation, P's are pump fluxes and Na^+, K^+, A^- are sodium, potassium and anions, respectively. (The J's labeled 1 are from compartment 1 to 2, 3 are from compartment 3 to 4, and unlabeled are across the membrane).

Changes in the bound cations are given by:

$$[Na^+]22 = k_f * Q2 * [Na^+]2 - k_r * [Na^+]22 \tag{7}$$

$$[K^+]22 = k_f * Q2 * [K^+]2 - k_r * [K^+]22 \tag{8}$$

$$[Na^+]33 = k_f * Q3 * [Na^+]3 - k_r * [Na^+]33 \tag{9}$$

$$[K^+]33 = k_f * Q3 * [K^+]3 - k_r * [K^+]33 \tag{10}$$

where $Q2$ and $Q3$ are the total surface charge, and the ratio of the binding and dissociation kinetic constants, $k_f/k_r = K_{eq}$, the binding equilibrium constant.

Changes in the negative surface charge on the membrane are given by:

$$\dot{Q}2 = - ([\dot{N}a^+]22 + [\dot{K}^+]22) - JQ \qquad (11)$$

$$\dot{Q}3 = - ([\dot{N}a^+]33 + [\dot{K}^+]33) + JQ \qquad (12)$$

(Equations 11 and 12 are split to allow for a fast moving fraction of charge and a slow moving fraction, so that gating current can be included).

Finally, the currents in the surface compartments and the membrane during voltage clamp [(dU4/dt) = 0] where U stands for potential, are given by:

$$I = F*(JNa^+1 + JK^+1 - JA^-1) - C1*(dU2/dt) \qquad (13)$$

$$I = F*(JNa^+3 + JK^+3 - JA^-3) + C3*[(dU3/dt) - (dU4)/dt] \qquad (14)$$

$$I = F*(JNa^+ + JK^+ - JA^- - JQ + PNa^+ + PK^+ - PA^-) + $$
$$+ C2*[(dU2/dt) - (dU3/dt)] \qquad (15)$$

where F is the Faraday constant and C1, C2 and C3 are the capacitances in the respective compartments. (We have retained the (dU4/dt) term because it is necessary for the initial current upon imposing the clamp). Equations 13-15 can be solved for derivatives of the U2 and U3 as well as I.

To study the behavior of these equations (BLANK, KAVANAUGH and CERF, [11]) it is necessary to set the initial conditions based on published values for the squid axon, *i.e.* the values of $[Na^+]_i$, $[K^+]_i$, $[A^-]_i$, Q_i and U_i. It is also necessary to establish ranges for the parameters, again based on experimental values when available, but including some simplifying assumptions. The parameters are:

1) two of the three binding parameters, the binding rate constant (k_f), the dissociation rate constant (k_r) and their ratio, the equilibrium constant (K_{eq}). These have been assumed to be the same for both sodium and potassium on the two sides of the membrane.

2) three ionic mobilities in the surface compartments that are all assumed to be equal (u) as a first approximation.

3) four conductances in the membrane, sodium (GNa^+), potassium (GK^+) anion (GA^-) and mobile charge associated with gating currents (GQ). The first three are fixed by the known steady state fluxes, and GQ is set according to the known behavior of the gating current.

4) three capacitances, two at the surfaces and one for the dielectric. These values are also fixed by observations, since the total capacitance must be about 0.8 $\mu F/cm^2$.

The equations were programmed on a digital computer for a voltage clamp experiment where $U4$ was changed from -65 mV to -20 mV *via* a very rapid (time constant ~ 20 μs) exponential function. (The program was written so that a small fraction of the surface charge would move rapidly across the membrane as a result of the depolarization, in line with the observed gating current). Many settings of the parameters were studied, but in all cases the mobilities (u) and conductances across the membrane (G) remained constant during a solution.

Results in one part of parameter space are shown in Fig. 9. At high values of GNa^+, GK^+ and low values of u, the depolarization produces a sodium current that goes in the positive direction. This unusual flux is due to the large changes in the chemical potential that result from cation accumulation and depletion on opposite sides of the membrane. (This occurs for both sodium and potassium, but in opposite directions and only the sodium flux is positive). The change in chemical potential more than compensates for the now unfavorable electrical driving force. In fact, the positive JNa^+ is greater at unfavorable electrical driving forces and at high values of k_r and K_{eq}.

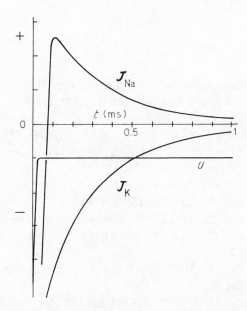

Fig. 9. — The net sodium flux, JNa^+, the net potassium flux, JK^+, and the voltage clamp potential, U, as functions of time when the membrane permeabilities to *both* sodium and potassium are raised several orders of magnitude. The transient sodium flux is positive (*i.e.* inward), while the potassium flux is negative. (Reproduced with permission from *Bioelectrochem. Bioenerg.*).

Bearing in mind that the model is approximate and that the results are far from complete, it appears that positive sodium fluxes can occur if there is a mechanism to increase the permeability as a result of depolarization. The permeability increase in this case is non-selective, more like the situation envisaged by BERNSTEIN. This raises interesting questions about the mechanism of the permeability increase, while at the same time suggesting that the system has certain physical properties (*e.g.* large ion binding equilibrium constants and fast ion release rate constants) as indicated by the magnitudes of the parameters.

From the above discussion it appears that unusual transient fluxes can occur when there are changes in the electrical double-layer regions at the surfaces of membranes. In the example given, the changes are due to imposed currents, but they can come about as result of many kinds of processes e.g. release of ions by the action of light (BLANK *et al.* [12]), redox reactions due to adsorbed enzymes. If the effects of electrical double-layers are taken into consideration in these cases, some of the unusual biological properties may turn out to be the physical consequences of specialized biological reactions at surfaces.

References

[1] S.E. LUX, *Nature* (*London*) **281**, 426 (1979).

[2] M. BLANK, *J. Colloid Interface Sci.* **75**, 435 (1980).

[3] M. BLANK, *Prog. Surf. Membr. Sci.* **13**, 87 (1979).

[4] M. BLANK and J.S. BRITTEN, *J. Colloid Sci.* **20**, 789 (1965).

[5] J.S. BRITTEN and M. BLANK, *Bioelectrochem. Bioenerg.* **4**, 209 (1977).

[6] M. BLANK, *Pharmacol. Ther.* **7**, 313 (1979).

[7] M. BLANK and S. FEIG, *Science* **141**, 1173 (1963).

[8] M. BLANK, *Colloids Surf.* **1**, 139 (1980).

[9] M. BLANK and J.S. BRITTEN, *Bioelectrochem. Bioenerg.* **5**, 528 (1978).

[10] M. BLANK and W.P. KAVANAUGH, *Bioelectrochem. Bioenerg.* **9**, (1982) in press.

[11] M. BLANK, W.P. KAVANAUGH and A. CERF, *Bioelectrochem. Bioenerg.* **9**, (1982) in press.

[12] M. BLANK, L. SOO, N.H. WASSERMAN and B. ERLANGER, *Science* **214**, 70 (1981).

Supplementary useful readings

J. BERNSTEIN, *Untersuchungen zur Thermodynamik der bioelektrischen Ströme;* *Pflügers Arch.* **92**, 521 (1902).

M. BLANK, *A Physical Interpretation of the Ionic Fluxes in Excitable Membranes;* *J. Colloid. Sci.* **20**, 933 (1965).

J.S. BRITTEN and M. BLANK, *The Effect of Surface Charge on Interfacial Ion Transport; Bioelectrochem. Bioenerg.* **4**, 209 (1977).

K.S. COLE, *Membranes, Ions and Impulses,* University of California Press, Berkeley (1968).

D.E. GOLDMAN, *Potential, Impedance and Rectification in Membranes; J. Gen. Physiol.* **27**, 37 (1943).

D.E. GOLDMAN, *Excitability Models in Biophysics and Physiology of Excitable Membranes,* ADELMAN (Editor), Van Nostrand-Reinhold, New York (1971), pp. 337-358.

P.T. GOULDEN, *The Biological Membrane Potential, Some Theoretical Considerations; J. Theor. Biol.* **58**, 425 (1976).

A.L. HODGKIN, *The Conduction of the Nervous Impulse,* Thomas, Springfield, Ill. (1964).

G. LING, *A Physical Theory of the Living State: The Association-Induction Hypothesis,* Blaisdell, New York (1962).

M.C. MACKEY, *Ion Transport through Biological Membranes; Lecture Notes in Biomathematics,* Springer Verlag, Berlin (1975), Vol. **7**.

J.Th.G. OVERBEEK, *Electrochemistry of the Double Layer* in *Colloid Science* R. KRUYT (Editor), Elsevier, Amsterdam (1952), Vol. **1**, pp. 115-193.

I. TASAKI, *Nerve Excitation: A Macromolecular Approach,* Thomas, Springfield, Ill. (1968).

T. TEORELL, *An Attempt to Formulate a Quantitative Theory of Membrane Permeability; Proc. Soc. Exp. Biol. Med.* **33**, 282 (1935).

T. TEORELL, *On Oscillatory Transport of Fluid Across Membranes; Acta Soc. Med. Upsalien.* **62**, 60 (1975).

THE MITOCHONDRIAL RESPIRATORY CHAIN

DAVID F. WILSON

Department of Biochemistry and Biophysics
University of Pennsylvania, Philadelphia, Pa. 19104, U.S.A.

Contents

1.   Introduction

The previous lecturers have discussed the properties of oxidation-reduction reactions. The purpose of this chapter is to describe the experimental techniques which have been utilized in studying the mitochondrial respiratory chain and our current ideas concerning the electrochemical behavior of oxidative phosphorylation. There are several important thermodynamic parameters which the experimentalyst needs to measure in order to attempt to describe an oxidation-reduction system. For any single oxidation-reduction component its reduction (or oxidation) can be written in the conventional form:

$$A_{Ox} + n\ e^- \longrightarrow A_{Red} \tag{1}$$

where the subscripts Ox and Red designate the oxidized and reduced species respectively of compound A. The equation describing the behavior of this component relative to a standardized hydrogen electrode is:

$$U_h = U_m - \frac{R\ T}{n\ F}\ \ln\left(\frac{[A_{Ox}]}{[A_{Red}]}\right) \tag{2}$$

With the proper experimental methods it is possible to measure the half-reduction potential $(U_m)^*$, the number of reducing equivalents accepted or donated per redox center and the number of moles of the compound present in a given volume of solution or suspension.

2.   Potentiometric measurements of U_m and n values

2.1.   Titrations with chemical oxidants and reductants

Potentiometric analysis as used in this laboratory has been described in detail in the literature [1-3]. I will use examples of the data obtained to indicate the techniques available and their applicability.

---

*   In classical electrochemistry the U_m value corresponds to the *standard* potential in case the quantities [A] are given in terms of activity, while it corresponds to the *formal standard* potential in case the same quantities are given in terms of concentrations.

2.1.1. Absorption measurements using visible light. – The *b* cytochrome of the mitochondrial respiratory chain had long been the subject of controversy, displaying anomalous kinetic and spectral behavior. In 1970 WILSON and DUTTON [4] observed that when suspensions of intact rat liver mitochondria were subjected to potentiometric titrations in the presence of ATP the *b* cytochrome reduction occurred in two $n = 1$ steps (Fig. 1), approximately half of the cytochrome had an $U_{m\,7.3}$ value of + 245 mV while the other half had an $U_{m\,7.3}$ value of + 35 mV. In the presence of uncoupler the two cytochromes were estimated to have U_m values of near 35 and − 55 mV although resolution was not achieved. These measurements

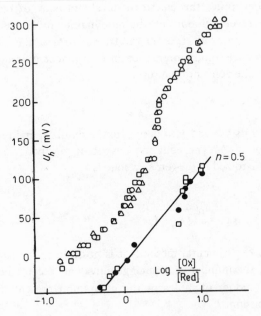

Fig. 1 A. – The dependence of the state of reduction of cytochrome *b* on the oxidation-reduction potential (U_h). The rat liver mitochondria were suspended at 0.25 μM cytochrome *a* in a medium containing 0.22 *M* mannitol, 0.05 *M* sucrose, 1 m*M* EDTA and 15 m*M* morpholinopropane sulfonate, pH 7.3. Rotenone 5 μM and the indicated redox mediators were added. Aliquots of ascorbate were added until anaerobiosis was achieved as evidenced by cytochrome reduction and an U_h of less than 300 mV. ATP 1.5 m*M* was added and the reduction of cytochrome *b* was effected by the combined slow donation of electrons from endegenous donors and the addition of small aliquots of DPNH. (△) 10 μM TMPD, 7 μM PMS, 4 μM PES and 50 μM FeSO$_4$. (□) 30 μM TMPD, 21 μM PMS, 12 μM PES and 50 μM FeSO$_4$. (○) 45 μM TMPD, 30 μM PMS, 9 μM PES and 50 μM FeSO$_4$. The points on the lower curve ($n = 0.5$) were for the conditions given above for the same symbol shape but measured after the addition of uncoupler [1 μM 5-Cl,3-*p*-phenyl,2′,4′5′-trichlorosalicylanilide].

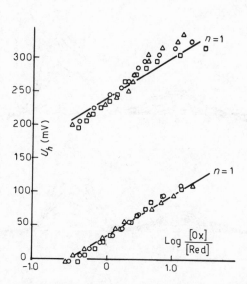

Fig. 1 B. — The oxidation-reduction potentials of the two components of cyto-chrome b in rat liver mitochondria in the presence of ATP. The logarithm of the ratio of the oxidized to reduced form of the component was calculated assuming 30 % the absorbance change was from the high potential component and 70 % was for the low potential component. The lines are for theoretical one electron accep-tors ($n = 1$) (taken from Ref. 4)

established the existence of two thermodynamically distinct b cytochromes and this was quickly followed by their spectral and kinetic characterization. ERECINSKA et. al. [5] used a scanning dual wavelength spectrophotometer in conjunction with potentiometric titrations of succinate-cytochrome c reductase to obtain elegant spectral resolution of the two b cytochromes (Fig. 2) and to show their different sensitivities to detergents. Thus it was possible to determine the absorption spectra of each b cytochrome in the absence of spectral interference from the other b cytochrome.

In uncoupled pigeon heart muscle mitochondria one b cytochrome had an $U_{m 7.2}$ value of $+ 30$ mV and the reduced form had a single α-absorption maximum at 561 nm (b_{561} or b_K) while the other b cytochrome had an $U_{m 7.2}$ value of $- 30$ mV and the reduced form had a split α-absorption with maxima at 565 nm and 558 nm (b_{565} or b_T), the latter appearing as a shoulder in room temperature spectra.

2.1.2. *Using electron paramagnetic resonance (e.p.r.) to measure the degree of reduction of the components*. — Many oxidation-reduction components cannot be measured using ordinary spectrophotometers because the absorbance change is too small to be readily measured or because of interference from other more strong-

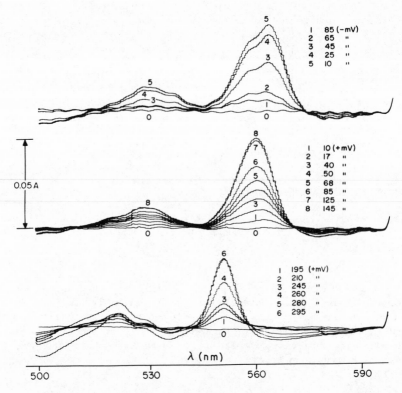

Fig. 2. — Difference spectra of the cytochromes of succinate-cytochrome c obtained during anaerobic potentiometric titration of the preparation. The succinate-cytochrome c reductase (2 mg protein/cm^3) was stirred under an ultrapure argon atmosphere in 0.1 M phosphate buffer pH 7.0 (0.008 % Triton X-100, 0.008 % DOC). The redox mediators used were: 20 μM diaminodurol, 40 μM each of phenazine methosulfate, phenazine ethosulfate, and duroquinone, 5 μM pyocyanine and 15 μM 2-hydroxy-1,4 napthoquinone. The figure shows an oxidative titration with potassium ferricyanide. The reference wavelength was 595 nm and the scanning speed was 1.3 min/100 nm. Upper traces represent the absorbance changes taken in the potential range: −100 mV to −10 mV; middle: −10 to +145 mV; bottom traces: +145 to +275 mV (taken from Ref. 5)

ly absorbing species. Some of these components have e.p.r. resonances which are greatly different for the oxidized and reduced species, permitting measurement of their state of reduction by this method. For most transition metals (such as iron and copper) compounds the measurements must be made on samples at low temperatures. A reaction vessel has been designed [6] which allows the oxidation-reduction potential of anaerobic samples to be adjusted to any desired value and then aliquots transferred to e.p.r. sample tubes under strictly anaerobic conditions. These samples can then be rapidly frozen using liquid isopentane at its freezing point

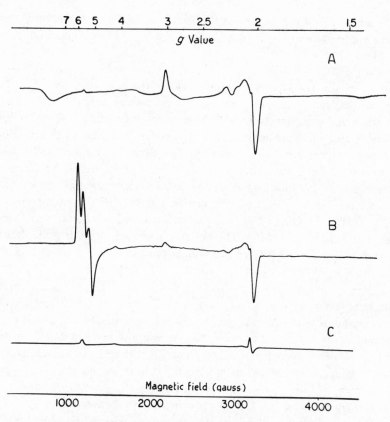

Fig. 3. — The oxidation-reduction potential dependence of the *e.p.r.* spectra of purified cytochrome *c* oxidase. The cytochrome *c* oxidase was dissolved at 75 μM heme *a* in a 30 mM phosphate buffer, pH 7.2. A sample of the aerobic preparation was frozen by immersion in *iso*pentane at its freezing point (spectrum A). The oxidation-reduction mediators phenazine methosulfate (60 μM), ferricyanide (~400 μM), and diaminodurene (100 μM) were added and aliquots taken at 300 mV (spectrum B) and 97 mV (spectrum C). The *e.p.r.* spectra were measured in a Varian E4 spectrometer with a modulation amplitude of 40 G and a power setting of 10 mW. The sample temperature was approximately 15 K (taken from Ref. 8)

(− 160 °C) and the *e.p.r.* absorbance measured at sample temperatures down to that of liquid helium and below. This technique was first used for measuring the half-reduction potential of mitochondrial iron-sulfur proteins [7] but has since been applied to most of the *e.p.r.* detectable components. Spectra from a titration of the components of cytochrome *c* oxidase by *e.p.r.* are shown in Fig. 3. In this experiment purified hemoprotein prepared by Dr. TSOO E. KING was used [8] in order to minimize interference from other parts of the respiratory chain. Aliquots of the suspension were frozen either aerobically (fully oxidized) or anaerobically at U_h values of 300 mV and 97 mV. Spectra of aerobic samples show resonances due to a copper atom (g = 2.00) and a low-spin ferric heme compound (g = 3.0, 2.2 and 1.5) are observed.

In partially reduced cytochrome oxidase the low-spin ferric heme signal has largely disappeared and strong resonances are observed near g = 6, typical of high-spin ferric heme compounds. Further lowering the potential to 97 mV fully reduces the enzyme and all of the heme and copper signals disappear as neither reduced iron nor reduced copper can be observed by $e.p.r.$ Similar titrations can be carried out using suspensions of mitochondria or submitochondrial particles, selecting the sample temperature and microwave power to optimize measurement of each component. Data from a typical potentiometric titration of the components of cytochrome oxidase in submitochondrial particles from pigeon breast muscle are shown in Fig. 4 (taken from Ref. 9). The amplitude of each signal is plotted against the oxidation-reduction potential of the sample prior to freezing. The g = 3 and g = 2 signals give n = 1.0 NERNST plots, with U_m values of 390 mV and 245 mV, respectively. When the absorbance changes of the heme components are measured in potentiometric titrations at room temperature a sigmoid curve characteristic of two heme compounds with different U_m values is observed. The calculated U_m for the higher and lower potential heme components are 380 mV and 210 mV respectively while that for the copper (measured by its absorbance change at 830 nm) is 245 mV. These values are within experimental error of those measured by low temperature $e.p.r.$ for the g = 3.0 and g = 2 species, indicating that freezing has not introduced a change in redox state of the components. A comprehensive survey of the U_m values for the respiratory chain components which have been measured by both methods, including cytochromes a, a_3, c, c_1, b_{565}, Cu and some iron sulfur proteins, reveals no significant differences ($\pm$ 10 mV) in any of the measured values and provides strong validation of the technique. One very interesting aspect of the $e.p.r.$ measurements is that the signals at g = 6, characteristic of high spin ferric heme is absent in both fully oxidized and fully reduced enzyme (Fig. 4). Its appearance corresponds to that expected for a component with an n value of 1.0 and an U_m of 390 mV.

Maximal intensity is observed near 300 mV and then the signal disappears with an n = 1.0, U_m = 190 mV. Since this signal is that of an oxidized heme quantitatively equal to one half of the total a heme [9], its appearance must be an expression of this component's interaction with another redox component whose reduction results in the appearance of the g = 6 signals. The U_m value for the appearance of the g = 6 signals should be the same as that for the component being reduced. The only component fulfilling this requirement is the other heme $i.e.$, the U_m for appearance of the g = 6 signals has the same value and pH dependence as does the disappearance of the g = 3 signal and the reduction of a heme as measured by the absorb-

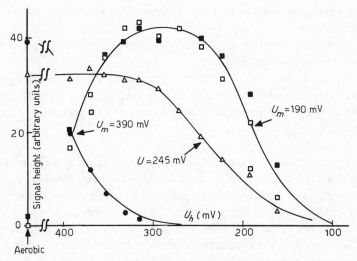

Fig. 4. – The oxidation-reduction potential dependence of the *e.p.r.* spectra of cytochrome *c* oxidase in submitochondrial particles. Pigeon breast muscle submitochondrial particles were suspended at 29 *μM* cytochrome *a* in a medium containing 0.2 *M* sucrose and 0.05 *M* morpholinopropane sulfonate, pH 7.0. Phenazine methosulfate (40 *μM*), diaminodurene (80 *μM*) and ferricyanide (100 *μM*) were added and anaerobic titrations carried out. The *e.p.r.* of the frozen samples at the temperature and power settings required for each signal. The signal amplitudes are plotted as a function of the U_h value of the sample before freezing. The signals represented are the *g* 6.4 maximum (□), the *g* 6.03 maximum (■), the *g* 3 maximum (●), and the copper signal (△) (taken from Ref. 9)

ance change at 445 nm or 605 nm in room temperature titrations. Only the high potential copper has a similar $U_{m\,7.0}$ value (350 mV) and it is pH-independent while that for the appearance of the $g = 6$ signals and for reduction of the high potential heme are pH-dependent. These results give clear evidence for the existence of strong interactions (heme-heme interaction) between cytochromes *a* and a_3, an interaction which has also been observed in their other measured properties (see for example Ref. 8-12).

3. Stoichiometry of oxidation-reduction components

3.1. Titration with chemical oxidant and reductant

Classically the number of moles of a component present in a sample has been measured in two ways:

1)   by measuring the amount of active component (Fe, quinone, etc.) by chemical analysis or

2)   by anaerobically titrating its oxidation and reduction with measured amounts of standard oxidant and reductant.

The apparatus designed for potentiometric analysis is also appropriate for anaerobic titrations of stoichiometry by the latter method. In general this technique is best suited to purified preparations as the presence of endogenous electron donors should be avoided.

In the cytochrome b-c_1 region of the respiratory chain cytochromes (c_1 and total b) can be analyzed chemically because the hemes are well characterized but the other redox components can not be as readily analyzed. Titration of purified cytochrome b-c_1 complex with chemical reductant (NADH) and oxidant (ferricyanide) has been utilized to measure the redox components of this part of the respiratory chain [12]. Reduction of the components can be monitored using the heme absorbance changes and a titration following the difference in absorbance between the wavelength pair 565-552 nm is shown in Fig. 5. These titrations fulfill the criteria for equilibrium as additions of oxidant or reductant induce rapid changes in absorbance which are then essentially time-independent and the oxidative and reductive titrations are in good agreement. When the absorbance change is replotted against

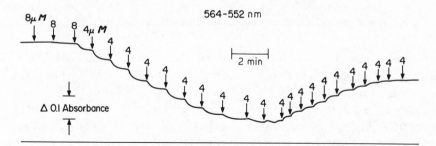

Fig. 5. — Spectrophotometric recording of the titration of an anerobic reduced preparation of the cytochrome b-c_1 complex with ferricyanide. The cytochrome b-c_1 complex was suspended at 15.4 μM heme c in 50 mM phosphate buffer, pH 7.2 and preequilibrated with ultrapure argon gas to remove most of the dissolved oxygen. Suffcent dithionite solution was added to reduce the complex and remove the remaining oxygen. The re-oxidation was carried out by stepwise addition of ferricyanide (the concentrations are given in the figure). Any excess of dithionite was destroyed by the initial addition of ferricyanide. Therafter, at least three oxidative (with ferricyanide) and reductive (with NADH; Fig. 6) titrations were carried out on each preparation. Thus there is no dithionite remaining in the incubation mixture at any given time during the titrations. The measuring wavelengths were 565-552 nm (taken from Ref. 12)

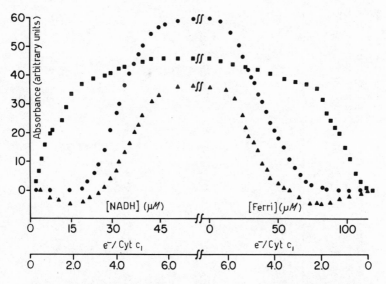

Fig. 6. – A plot of an oxidative (ferricyanide) and reductive (NADH) titration of the cytochromes of b-c_1 complex. The experimental data were obtained from titrations similar to those shown in Fig. 5: (■), absorbance changes at 554 nm; (●), absorbance changes at 560 nm; (▲), absorbance changes at 566 nm. The heme c concentration was 16.9 μM (taken from Ref. 12)

the number of reducing equivalents added or subtracted per mole of cythochrome c_1 (Fig. 6) three conclusion are possible:

1)   two equivalents per c_1 are required for reduction of cytochrome c_1;
2)   three equivalents per cytochrome c_1 are required to reduce the b cytochromes;
3)   a careful summation of the known redox components indicates that these account for all of the reducing equivalents measured in the titrations (Table 1).

Spectra of the sample at each point in the titrations identifies the heme components being oxidized or reduced. Cytochrome b_{561} is reduced before cytochrome b_{565} and computer fitting of the curves using the measured half-reduction potentials and amounts of the known components show that a good fit requires equal concentrations of the two b cytochromes, a result difficult to obtain by other methods. Even more important, the measurements eliminate the possibility that other, as yet unidentified, redox components are present in the cytochrome bc_1 complex at stoichiometries greater than approximately 0.3 times that of the cytochrome c_1.

Table 1. — Electron acceptors of the b-c_1 complex of the respiratory chain and their titrations with NADH and ferricyanide

Preparation number	Heme c (µM)	NADH		Ferricyanide		Number of electron equivalents (mM) in preparation						
		µM	e^-/c_1	µM	e^-/c_1	Heme c	Heme b	NHI[a]	Ubiqui-none[b]	SDH[c]	Total	e^-/c_1
1	14.4			67.5	4.67	0.12	0.226	0.115	0.097	0.0486	0.608	5.07
	16.8			73.5	4.38							
Average					4.53							
2	17.8	48.7	5.47	110	6.2	0.148	0.258	0.135	0.102	0.044	0.687	4.64
	17.8	46.5	5.22	100	5.65							
	14.8	35.0	4.7	75	5.1							
	14.8	32.5	4.4	77.5	5.2							
Average			4.94		5.54							
3	11.0	28.0	5.1	67.5	6.1	0.11	0.205	0.105	0.082	0.049	0.581	5.01
	15.4	37.5	4.9	75.0	4.9							
Average			5.0		5.5							
Average for three preparations			4.97 ± 0.32		5.25 ± 0.61							4.91

[a] (Heme c + heme b)/3

[b] A two electron donor/acceptor

[c] Flavin (2 e^-) + NHI (2 e^-)

3.2.   Titration using electrical currents

One of the recent developments, at least for biological systems, is the use of polarized electrodes to inject or remove electrons from the system. Biological compounds, in particular lipids and some proteins, have a tendency to coat and thereby inactivate electrode surfaces. Recently electrometric reaction systems have been designed which can be used to titrate purified preparations in detergent suspension [13] and even membraneous preparations in the absence of detergents [14]. A four electrode cuvette, capable of electrically providing the current required to oxidize and reduce anaerobic suspensions of submitochondrial particles such as to obtain either the coulombs of electricity required to reduce the redox components or for controlled potentiometric titration of these compounds, is shown in Fig. 7. This cuvette has a vitreous carbon working electrode of large surface area (0.8 cm diameter) mounted in the bottom where its surface is continuosly cleaned by the magnetic stirring bar which runs on its surface. The counter electrode (Ag/AgCl) is in a side chamber connected by a low resistence Agar salt bridge and the circuit utilizing the working and counter electrodes is electrically isolated from the measuring electrode circuit. The measuring electrodes consist of a piece of platinum foil attached to the inside of the reaction chamber and a calomel electrode connected to the reaction vessel via a small flexible plastic tube filled with 5 % Agar in 1 M KCl. The cuvette may be used either to determine the coulombs of electricity required to oxidize and reduce the redox components in the sample or as a potentiometric cuvette for which the electrical circuit can be set to bring the measuring electrodes to any desired voltage reading.

3.2.1. Coulometric measurement of the stoiochiometry of the oxidation-reduction components of the cytochrome c oxidase region of the respiratory chain.

– Suspension of submitochondrial particles from pigeon breast muscle were placed in the electrometric cuvette and 0.8 mM methyl viologen ($U_{m\,7.0}$ = 0.44 V) added. Reduced methyl viologen was generated electrometrically to exhaust residual oxygen and reduce the respiratory chain components. Ferrocene ($U_{m\,7.0}$ ≈ 0.56 V) was then added and the particles subjected to cycles of oxidation and reduction. The absorbance changes of the hemes of cytochrome oxidase were measured at 605 - 630 nm as electrons were injected into or removed from the sample in the cuvette.

The oxidative and reductive titrations are very similar and quantitative comparison is readily achieved by plotting the change in absorbance against the coulombs of electricity injected or removed from the reaction medium (Fig. 8). The data are given for titrations carried out using three different protein concentrations in order to indicate the precision of the method.

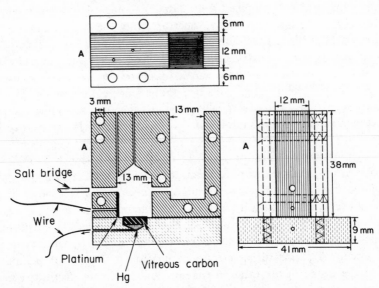

Fig. 7. — A four electrode coulometric-potentiometric titration cuvette. The cuvette is constructed of plexiglas with the central portion, from which the reaction chamber is made, of black plexiglas to prevent light from passing through any portion except the reaction chamber. The reaction chamber has a volume of 1.79 cm^3 and a light path of 1.2 cm. The top of the chamber is conical with a 1.2 mm diameter hole extending 2 cm from its apex to the top of the cuvette. Solutions are added and withdrawn through a second hole 1.0 mm in diameter which is placed near the edge of the chamber. When a solution is added through the smaller hole using a syringe with a 25 gage, 1 1/2″ needle all of the air, including bubbles, is forced out through the central hole. Filled to the top each access hole contains a 2 cm column of unstirred liquid, an effective barrier to oxygen diffusion.

The working electrodes consist of a 0.8 cm diameter button of polished vitreous carbon mounted in the bottom of the reaction chamber and a Ag/AgCl electrode placed in a 1 M KCl solution in the open chamber adjacent (0.8 cm separation) to the reaction chamber. Electrical connection is through a 3 mm diameter hole 1 cm long between the chambers which is normally filled with a 4 % Agar gel containing 1 M KCl.

The sensing electrodes consist of a small piece (0.6 cm^2) of platinum foil on the side of the reaction chamber and an external calomel electrode. A flexible plastic tube (1.5 mm i.d.) filled with 4 % Agar containing 1 M KCl is used to provide electrical contact between the reaction medium and the external calomel electrode.

The reaction mixture is stirred using a 1 cm $\times$ 2 mm teflon coated magnetic stirring bar and an external magnetic stirrer. The stirring bar runs on the surface of the vitreous carbon electrode (from Ref. 14)

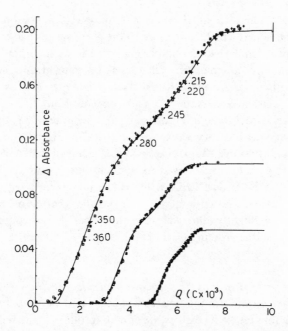

Fig. 8. — Coulometric titration of cytochrome c oxidase in intact mitochondrial membranes. Pigeon breast muscle submitochondrial particles were suspended at approximately 6, 3, or 1.5 mg protein per cm^3 in a 0.1 M phosphate buffer, pH 7.0. Methylviologen (1.1 mM), anthroquinone-2,6-disulfonate (0.67 mM), phenazine ethosulfate (11 μM) and ferrocene (55 μM) were used to catalyze electron transfer at the working electrode. Reduction and oxidation of cytochromes a and a_3 was measured at 605 nm minus 630 nm. At each protein concentration both oxidative and reductive titrations were carried out at working electrode currents between 15 and 40 μA. The solid lines have the expected behavior assuming the presence of 6 redox components (cyt. $a + a_3$, high potential copper, low potential copper, cytochrome c_1, and the RIESKE iron-sulfur protein) of equal stoichiometry and $U_{m7.0}$ values of 0.220 V, 0.350 V, 0.360 V, 0.245 V, 0.215 V, and 0.280 V, respectively. The calculated extinction coefficients for cytochromes a_3 and a are 15.3 mM^{-1} cm^{-1} and 11.1 mM^{-1} cm^{-1}, respectively for a combined value of 26.4 mM^{-1} cm^{-1} (taken from Ref. 14)

Each titration curve shows three phases, the first (high potential) portion accounts for approximately 50 % of the absorbance change and one third of the coulombs. This relative stoichiometry is consistent with the high potential phase representing cytochrome a_3 and the high potential copper while the low potential phases account for the remaining two thirds of the coulombs and include cytochrome a and low potential or *visible* copper in addition to the RIESKE iron-sulfur protein and cytochrome c_1.

Quantitative analysis of the curves requires consideration of the effect of reduction of the hemes which contribute to the absorbance change as well as that of the components which have either no or small contributions to the measured absorbance change but are stoichiometrically equal in concentration to the heme. In Fig. 8 theoretical curves are fitted to each set of experimental data. These theoretical curves are constructed assuming equal concentrations of cytochrome a_3, high potential copper, cytochrome a, cytochrome c_1, low potential copper, and RIESKE iron-sulfur protein with respective U_m values of 360 mV, 350 mV, 215 mV, 220 mV, 245 mV, and 280 mV. A precise fit to the data is obtained when an extinction coefficient of 26.4 $mM^{-1} cm^{-1}$ is used for cytochromes a and a_3, in agreement with the value reported by VAN GELDER [15]. The non-linearity of the low potential phase is consistent with potentiometric data showing that cytochrome a has a half-reduction potential slightly more negative than does the low potential copper and the RIESKE iron-sulfur protein [11, 16].

3.2.2. Potentiometric titrations of cytochrome oxidase in the presence and absence of CO. – The electrometric method can equally well be used for potentiometric titrations if suitable redox mediators are added. In this case the redox mediators are present at concentrations higher than that of cytochrome oxidase and the working electrode functions primarily to reduce or oxidize the mediators. The platinum sensing electrode equilibrates with the oxidation-reduction mediators in the reaction medium allowing the oxidation-reduction potential to be continuously measured. The regulatory electrical circuits can be set to cause the working electrode to attain and hold to within ±1 mV any desired voltage on the sensing electrodes. Thus potentiometric titrations can be carried out using precise increments in oxidation-reduction potential.

A typical titration at pH 8.0 and in the presence of 50 μM CO is shown in Fig. 9 A, where steps of 20 mV were used for both reductive and oxidative titrations with the potential relative to a saturated calomel electrode given for each step. In Fig. 9 B the data are presented as the logarithm of the ratio of oxidized to reduced forms plotted against the oxidation-reduction potential relative to a standard hydrogen electrode. The curve is resolved into a component with an n value of 1.0 (cytochrome a) and a component with an n value of 2.0 (cytochrome a_3-CO compound). These components have $U_{m\,8.0}$ values of 385 mV (n = 2.0) and 245 mv (n = 1.0) respectively (see also Ref. 17).

Spectra measured for each step change in oxidation-reduction potential clearly show the formation of the CO compound with its absorption maximum at 589 nm and reduction of cytochrome a with an absorption maximum at 605 nm.

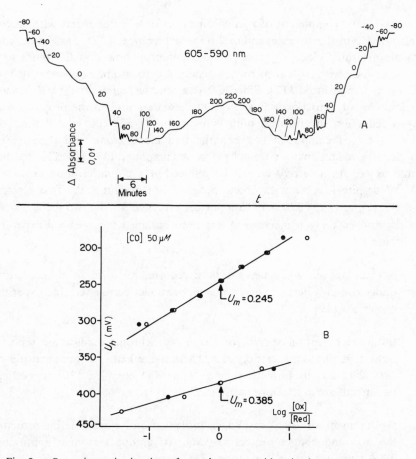

Fig. 9. — Potentiometric titration of cytochrome c oxidase in the presence of CO. Submitochondrial particles were suspended at approximately 6 mg protein per cm³ in 0.1 M phosphate buffer, pH 8.0. Methyl viologen (1.1 mM) was added, and then the working electrode circuit polarized. After anaerobiosis was attained, ferrocene (0.5 mM), ferricyanide (0.2 mM), diaminodurene (60 μM), and phenazine methosulfate (100 μM) were added. Titrations were carried out measuring the absorbance at 605 - 590 nm and the regulating electrical circuit was set to use the working electrodes to bring the sensing electrodes to designated voltages relative to the standard calomel electrode. In A the data are shown as the measured absorbance change as a function of time with the potential reading on the sensing electrode when $dA/dt = 0$ indicated at the appropriate places on the tracing. In B these data are plotted according to the NERNST equation and lines drawn for best fit to the behavior of a one electron acceptor (cytochrome a) or two electron acceptor (cytochrome a_3-CO compound) (taken from Ref. 14)

4. Electrochemical reduction of oxygen to water by cytochrome c oxidase

The heme component of cytochrome c oxidase which reacts with CO has an n value of 1.0 in the absence, and of 2.0 in the presence of CO. These data, coupled with the measured electron acceptor stoichiometry, show that CO binds to cytochrome oxidase with highest affinity when both cytochrome a_3 and the high potential copper are reduced [17]. Since CO is a strictly competitive (with respect to O_2) inhibitor of mitochondrial respiration these data lead to the proposal that the oxygen reduction site involves both cytochrome a_3 and copper and that oxygen forms a bridged compound between the two metal atoms and then undergoes two electron reduction to a bound peroxide compound [17, 18]. This mechanism is able to explain how oxygen can be reduced at a potential more positive than 600 mV despite its very unfavorable potential for accepting the first electron to form O_2^- ($U_{m\,7.0} \sim -320$ mV). Supporting evidence for the two metal atom reaction site-bridged oxygen mechanism has been obtained using several experimental approaches.

1) NO, an oxygen analogue, has been reported to form an *e.p.r.* identifiable bridge complex between heme and copper under certain selected experimental conditions [19].

2) Infrared absorption spectra of CO in cytochrome oxidase are reported to show that photodissociation of CO (from the heme) at temperatures below 140 K results in formation of a copper-CO complex [20] suggesting the two metals are in the same active site.

3) Measurement of the X-ray edge absorption and extended fine structure of the iron and copper centers suggests that a copper atom is approximately 3.7 Å from one of the iron atoms [21].

The existence of a bridged peroxide intermediate in the reaction with oxygen has not yet been unequivocally established. Analysis of the steady state rate of oxidation of reduced cytochrome c in phosphorylating mitochondria shows the existence of an intermediate with the expected kinetic behavior [18, 22]. Pre-steady state kinetic measurements of the reaction of molecular oxygen with isolated oxidase and mitochondria at low temperatures [24-27] show at least three identifiable intermediates in the reaction. It has proven impossible as yet to unambiguously assign structures to these intermediates or to be sure of their relevance to the physiological reaction.

5. Integrated function of the respiratory chain

The known oxidation-reduction components of the mitochondrial respiratory chain are listed in Table 2, together with their half-reduction potentials at pH 7.2 (in uncoupled mitochondria), approximate concentration relative to cytochrome a and the pH dependence of the half-reduction potential.

An examination of the half-reduction potentials of the redox components of the respiratory chain shows that the half-reduction potentials of the components fall into four groups. The most negative group is associated with NADH dehydrogenase and is matched with the U_m of the NAD couple at near -0.30 V. The next more positive group of components is associated with the cytochrome b-ubiquinone region of the respiratory chain and succinate dehydrogenase and is near 0.00 V. The third group includes the cytochrome c-cytochrome a region of the respiratory chain (U_m near 0.24 V), while the fourth *group* is not really a *group* but includes the oxygen active site which utilizes cytochrome a_3 and the high potential copper with U_m values near 0.36 V. In suspensions of coupled mitochondria carrying out net ATP synthesis the components within each of the first three groups behave as equilibrated pools of reducing equivalents in which they have, within experimental error, the same U_h value. The pools have U_h values near -0.31 V, 0.00 V, 0.31 V and 0.63 V [31, 32]. It is transfer of reducing equivalents between the pools which is energetically coupled to the phosphorylation of ADP. All three phosphorylation sites are coupled to the same pool of [ATP]/[ADP][Pi] through a common non-phosphorylated *high energy* intermediate [33, 34]. This sharing of common reactants, and the fact that reducing equivalents from NADH must pass sequentially through all three phosphorylation sites, assure that the same change in oxidation-reduction potential occurs for each of the sites in well coupled mitochondria.

To perform precise analysis of the free energy changes associated with the redox reactions of the respiratory chain, it is necessary to know the U_h values of redox components on the reducing and oxidized side of the site(s). In our own work we prefer to determine ΔU for two phosphorylation sites by utilizing the intramitochondrial $NAD^+/NADH$ couple on the reducing side and cytochrome c on the oxidizing side. In suspensions of isolated mitochondria the redox state of the NAD couple can be measured by the absorbance and fluorescence changes that occur on reduction and oxidation of the intramitochondrial NAD couple or by enzymatic analysis of the substrate couples which enzymatically equilibrate with the intramitochondrial NAD couple. The former methods are relatively unspecific because the intramitochondrial NADPH also contributes to the absorbance and fluorescence changes. In addition, of the available methods only analysis of the appropriate substrate couples gives U_h values which are not subject to errors due to binding

Table 2. – The oxidation-reduction components of the mitochondrial respiratory chain

Respiratory chain fragment	Component	Approximate relative conc.	$U_{m\,7.2}$	pH dependence
NADH dehydrogenase	FMN	0.15	−0.380	60 mV
	(Fe-S) N1a	0.15	−0.020	60 mV
	(Fe-S) N-2	0.15	−0.240	0 mV (6-8.6)
	(Fe-S) N-3	0.15		
Succinate Dehydrogenase	FAD	0.3	−0.040	?
	(Fe-S) S-1	0.3	0.030	0 mV
	(Fe-S) S-3	0.3	0.060	?
	Ubiquinone 10	10-15	0.045	60 mV
Cytochrome b-c_1 Complex	Cyto. b_{561}	1.0	0.030	60 mV
	Cyto. b_{566}	1.0	−0.030	60 mV
	Cyto. c_1	1.0	0.215	0 mV
	Rieske Fe-S	1.0	0.280	0 mV
Cytochrome c	Cyto. c	2.0	0.235	0 mV
Cytochrome c Oxidase	Cyto. a	1.0	0.210	20 mV
	Low potential Cu	1.0	0.245	0 mV
	Cyto. a_3	1.0	0.385	{ 0 mV (< pH 7) 60 mV (> pH 7)
	High potential Cu	1.0	0.350	0 mV

For additional discussion of the individual values and original references see a review by Ohnishi [1] on the iron-sulfur proteins and general and general reviews by Dutton and Wilson [2] and Wilson [3].

of the NAD^+ and NADH to the mitochondrial content. In intact cells use must be made of the equilibration of the intramitochondrial NAD couple with various substrate couples for which the enzyme is present only in the mitochondrial matrix (such as 3-hydroxybutyrate dehydrogenase or glutamate dehydrogenase).

On the oxidizing side of the second phosphorylation site, cytochrome c is the best choice because it has readily measured spectral changes upon oxidation and reduction, its U_m value is accurately known and its U_m value is independent of both pH and [ATP]/[ADP][Pi]. Two phosphorylation sites are associated with the transfer of reducing equivalents from NADH to cytochrome c. The equation for the coupled reaction is

$$NADH_m + 2\ c^{3+} + 2\ ADP_c + 2\ Pi_c \rightleftharpoons NAD_m + 2\ c^{2+} + 2\ ATP_c \qquad (3)$$

Experimentally it has been shown that mitochondrial oxidative phosphorylation is regulated by and equilibrates with the cytosolic (extramitochondrial) [ATP], [ADP], and [Pi] [35, 36, 40]; this is indicated in equation (3) by the subscript c associated with these reactants while the subscript m is used to indicate that the intramitochondrial NAD and NADH are the species which donate the reducing equivalents to the respiratory chain.

5.1. Coupling of the oxidation-reduction reactions to ATP synthesis: free enthalpy* relationships

The free enthalpy of synthesis (or hydrolysis) of ATP can be readily analyzed.

Given the reaction

$$ATP \longrightarrow ADP + Pi \qquad (4)$$

the free enthalpy change (ΔG_{ATP}) may be calculated from

$$\Delta G^{\circ\prime}_{ATP} = \Delta G_{ATP} + RT \ln \frac{[ADP][Pi]}{[ATP]} \qquad (5)$$

The $\Delta G^{\circ\prime}_{ATP}$ has been measured for the relevant experimental conditions [41, 42] and measurements of the concentrations of ATP, ADP, and phosphate

---

*   To avoid misunderstanding it is reminded here that the *free enthalpy* ΔG is the more logical name of the (GIBBS) *free energy* and that this name is more and more gaining favour in physiological chemistry.

under various conditions allow calculation of ΔG_{ATP}. This means that the free enthalpy change for the oxidation-reduction reactions ($\Delta G_{Ox\text{-}Red}$) as well as that for ATP synthesis (ΔG_{ATP}) can be measured with reasonable accuracy.

In Table 3 measurements of the free enthalpy changes of the transfer of reducing equivalents from NADH to cytochrome c are presented as the measured U_h values for the NAD couple and cytochrome c, the ΔU for the redox reactions, and the free enthalpy change for the transfer of two reducing equivalents [37-40, 43, 44] from NADH to cytochrome c. The U_h of the NAD couple ranges from -0.34 V in suspensions of isolated mitochondria to -0.24 V in P. denitrificans and T. pyriformis cells, while the U_h of cytochrome c is near 0.27 V. The ΔU values of 0.61 V and 0.52 V correspond to $\Delta G_{Ox\text{-}Red}$ values of -28.4 kcal/2 eq (-119 kJ/2 eq) to -24.0 kcal/2 eq (-100 kJ/2 eq).

The free enthalpy of ATP hydrolysis has been determined in each case by measuring either the extramitochondrial or whole cell [ATP], [ADP], and [Pi] and calculating the free energy change using the appropriate $\Delta G^{o\prime}$ for ATP hydrolysis (Table 3). This value was then used to calculated the ΔG for synthesis of two moles of ATP (ΔG_{ATP}) as required for comparison with the transfer of reducing equivalents from NADH to cytochrome c. The data show that within experimental error the coupled reactions are near equilibrium (ΔG is near zero). Thus any proposed intermediate in oxidative phosphorylation must also be near equilibrium with both the oxidation-reduction reactions and the [ATP]/[ADP][Pi].

The question of whether the mitochondrial respiratory chain should be considered to react with and be regulated by the cytosolic or the mitochondrial adenine nucleotides has remained controversial (compare Refs. 35, 36, 40 with Refs. 45-48).

Near equilibrium of reaction (3) suggests that whatever the role of the intra-mitochondrial [ATP], [ADP], and [Pi] and the adenine nucleotide translocase, the reactions occur without changing the overall stoichiometry of oxidative phosphorylation. There are several lines of evidence which support the idea that the cytosolic [ATP], [ADP], and [Pi] are the important species in regulation of mitochondrial respiration. The evidence is in five parts:

1.   Measurements of the stoichiometry of the synthesis of extramitochondrial ATP from NADH to oxygen gives approximately 3 ATP per NADH [32-35], equal to the number of transduction sites present in the respiratory chain.

2.   Measurements of the regulation of respiration in suspensions of isolated mitochondria show a dependence on the extramitochondrial [ATP]/[ADP] and

Table 3. — The free enthalpy relationships between the oxidation-reduction reactions of the respiratory chain and ATP synthesis[a]

Material	$U_{h\,NAD}$ (V)	$U_{h\,C}$ (V)	ΔG_{ATP}				$\Delta\Delta G$		Ref.
			kJ/2 eq	(kcal/2 eq)	kJ/mol	(kcal/2mols)	kJ	(kcal)	
Pigeon heart mitochondria	−0.343	0.270	−119	(−28.4)	−124.7	(−29.8)	5.7	(1.4)	10
Pigeon heart mitochondria	−0.308	0.312	−119.7	(−28.6)	−121.3	(−29.0)	1.6	(0.4)	10
Rat liver cells	−0.260	0.269	−102	(−24.4)	−101.3	(−24.2)	−0.7	(−0.2)	11
Ascites tumor cells	−0.270	0.260	−102	(−24.4)	−98.7	(−23.6)	−3.3	(−0.8)	12
Cultured kidney cells	−0.252	0.271	−100.8	(−24.1)	−101.3	(−24.2)	0.4	(0.1)	16
Paracoccus denitrificans	−0.244	0.276	−100.4	(−24.0)	−100.8	(−24.1)	0.4	(0.1)	17
Tetrahymena pyriformis	−0.236	0.251	−94.1	(−22.5)	−93.7	(−22.4)	−0.4	(−0.1)	13

[a] $U_{h\,NAD}$ and $U_{h\,C}$ are the oxidation-reduction potentials of the NAD couple and the cytochrome c couple respectively, while $\Delta G_{Ox\text{-}Red}$ is the free enthalpy change accompanying transfer of two reducing equivalents from NADH to cytochrome c. The $\Delta\Delta G$ is calculated as $\Delta G_{Ox\text{-}Red}$ minus ΔG_{ATP} and under the experimental conditions must be negative. The accuracy of the measurements is approximately ± 4 kJ (± 1 kcal) except in the case of isolated mitochondria where the accuracy with which the potential of the NAD couple can be measured is less [37].

[Pi] which is consistent with a dependence on the free enthalpy of synthesis of ATP, that is, [ATP]/[ADP][Pi] ([35, 36], see, however, Refs. 45-48).

3.   In experiments in which the cellular [Pi] was changed by metabolic methods (glycerol or fructose addition) or by changing the extracellular [Pi], the cellular [ATP]/[ADP][Pi] remained constant when the respiratory rate was constant [49].

4.   In ascites tumor cells [50] respiration and glycolysis increase in parallel to the increased rate of ATP utilization caused by glucose addition and then decrease again in parallel when the extra ATP utilization ends and the phase of inhibited respiration is entered; this is called the *Crabtree effect*. Again [Pi] is as important as [ATP] and [ADP].

5.   Experiments using the prokaryotic bacterium *P. denitrificans* [49] show that the same thermodynamic relationship between the respiration rate and cellular [ATP]/[ADP][Pi] is observed as in eukaryotic organisms with mitochondria, despite the fact that the former has only a single intracellular compartment (no requirement for adenine nucleotide translocase).

5.2.   Regulation of the rate of mitochondrial respiration and ATP synthesis

As elegantly pointed out by KREBS and VEECH [51], metabolic pathways consist of reaction networks in which some of the reactions are very near thermodynamic equilibrium and others are essentially irreversible. The near-equilibrium reactions allow for high thermodynamic efficiency; the essentially irreversible reactions serve to regulate flux through the pathway. In the previous section evidence has been presented that oxidative phosphorylation from NADH to cytochrome c is near equilibrium, giving rise to a high efficiency (nearly 100 %) for the transduction of electrochemical energy in this part of the respiratory chain to ATP synthesis. The final phosphorylation site in the respiratory chain occurs at cytochrome c oxidase and involves the overall reaction.

$$ADP_c + Pi_c + 2 c^{2+} + 1/2 O_2 + 2H^+ \longrightarrow 2 c^{3+} + H_2O + ATP_c \qquad (6)$$

The reducing equivalents are transferred from cytochrome c (U_h = 0.25-0.30 V) to molecular oxygen (U_h = 0.82 V) and this transfer occurs with a negative free enthalpy change of 100-109 kJ/2 eq of which only 50-67 kJ is utilized to synthesize ATP. The remaining negative free energy change is lost as heat, making the reaction highly irreversible. It is therefore this reaction which is responsible for the regulation of the rate of mitochondria phosphorylation [31, 32, 35, 49, 50].

5.3. Dependence on the state of cytochrome c reduction

In order to measure the dependence of mitochondrial respiratory rate on cytochrome *c* reduction it is necessary to utilize respiratory substrates which react directly with cytochrome *c*. The redox dye N, N, N', N'-tetramethyl-*para*phenylene-diamine (TMPD) catalyzes the reduction of cytochrome *c* in intact mitochondria by ascorbate and in the presence of excess (10 mM) ascorbate the steady state level of reduction of cytochrome *c* increases as the TMPD concentration is increased [52]. Simultaneous measurements have been made of the reduction of cytochrome *c* and respiratory rate of suspensions of mitochondria in both low energy states (+ uncoupler or ADP + Pi) and high energy state (+ ATP). When the [ATP]/[ADP] [Pi] is less than approximately 10^2 M^{-1} the respiratory rate is proportional to the % reduction of cytochrome *c* (Fig. 10). As the [ATP]/[ADP][Pi] is increased however the respiratory rate decreases and becomes a non-linear function of the reduction of cytochrome *c*. It should be noted that although TMPD was used to adjust the level of cytochrome *c* reduction, the measured respiratory rate is the same for all substrates if the cytochrome *c* reduction, [ATP]/[ADP][Pi] and pH are the same. Computer fitting of this data to the kinetic equations to the proposed model for oxygen reduction by cytochrome oxidase [35] have been carried out. These show that the observed regulation of the cytochrome $c \longrightarrow O_2$ reaction is quantitatively consistent with this reaction being responsible for the control of mitochondrial respiration.

5.4. Dependence on [ATP], [ADP], [Pi] and the oxidized substrate used

Experimental measurements of the dependence of the rate of respiration of suspensions of isolated mitochondria on [ATP], [ADP] and [Pi] have been reported to show that [Pi] is as important as [ADP] and [ATP]. This is observed in Fig. 11 (taken from Ref. 9) where the respiratory rate is the same function of [ATP]/[ADP] [Pi] at [Pi] values from 1.6 mM to 8.5 mM. Other workers (see for example Refs. 45, 46, 48) have reported that the respiratory rate is [Pi] independent, characteristic of its being limited by the rate of adenine nucleotide translocation. This experimental difference appears to be the result of differences in the experimental conditions used (for a review see Ref. 54), in particular the Mg^{2+} concentration. *In vivo* the role of [Pi] is unambiguous [49] and there can be little doubt that the respiratory rate is dependent on [ATP]/[ADP][Pi].

Although cytochrome *c* oxidase determines the rate of transfer of reducing equivalents through reaction 4 and therefore the flux through oxidative phosphorylation, clearly the overall flux is dependent on the availability of reducing substrate for the respiratory chain (the intramitochondrial [NAD$^+$]/[NADH]) and the energy

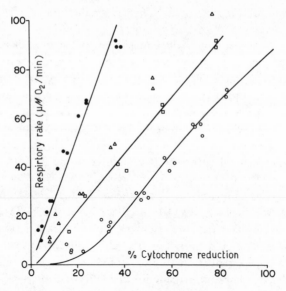

Fig. 10. — The dependence of the mitochondrial respiratory rate on the reduction of cytochrome *c* at pH 7.0. The experimental points represent data obtained by suspending pigeon heart mitochondria at approximately 0.2 μM cytochrome *a* in a medium containing 0.2 *M* sucrose, 0.04 *M* morpholinopropane sulfonate, and 0.2 m*M* ethylene dinitrilotetraacetate, pH 7.0. Sodium ascorbate (5 m*M*), ATP, ADP, and Pi were added and then aliquots of TMPD used to stimulate respiration. The respiratory rate and the reduction of cytochrome *c* (550 nm minus 540 nm) were measured simultaneously, and the resulting data were plotted in the figure. The added concentrations of ATP, ADP, and Pi were 4 m*M* ATP (○); 5 m*M* ATP, 1 m*M* ADP, and 1 m*M* Pi (□): 1 m*M* ATP, 1 m*M* ADP, and 1 m*M* Pi (△); and 3 μg of oligomycin plus 0.34 μM S-13 (●). Control measurements with 3 μg of oligomycin plus 0.023 μM S-13 in addition to the indicated ATP, ADP, and Pi were indistinguishable from those of just oligomycin and S-13. The solid curves are the simulated behavior of a suspension of mitochondria at a concentration of 0.4 μM cytochrome *c* (0.2 μM cytochrome *a*) and [ATP]/[ADP][Pi] values of 10^{-1} M^{-1}, 5×10^3 M^{-1} and 8×10^5 M^{-1} (taken from Ref. 25)

required for ATP synthesis, [ATP]/[ADP][Pi]. These parameters influence the flux through reaction 4 because through near equilibrium in reaction 1 they determine the state of reduction of cytochrome *c*:

$$\frac{[c^{2+}]}{[c^{3+}]} = \left(\frac{[NADH]}{[NAD^+]}\right)^{\frac{1}{2}} \frac{[ADP][Pi]}{[ATP]} K_{eq}^{\frac{1}{2}} \tag{7}$$

where K_{eq} stands for the equilibrium constant at a given pH.

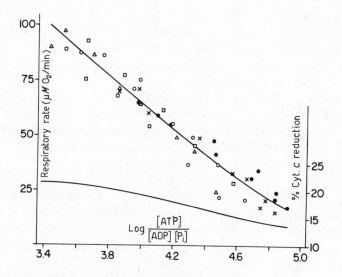

Fig. 11. — The dependence of the mitochondrial respiratory rate on [ATP], [ADP], and [Pi]. The simulated rate of mitochondrial respiration is plotted on the ordinate scale on left and the logarithm of the [ATP]/[ADP][Pi] value is plotted on the abscissa. The reduction of cytochrome c as obtained from the curvefitting process is also plotted on the ordinate (scale on right). The respiratory rate data for suspensions of dog heart mitochondria as taken from Ref. 9 are plotted for comparison. Each symbol is for data taken at a different phosphate concentration. The values are 1.6 ($\bullet$), 3.2 ($\times$), 5.1 ($\circ$), 7.0 ($\square$), and 8.5 mM ($\triangle$) (taken from Ref. 25)

In addition, the reaction of cytochrome c oxidase involves ADP phosphorylation and is itself dependent on [ATP]/[ADP][Pi] [53].

In vivo the reduction of cytochrome c may be regarded as a dependent variable, dependent on [NADH]/[NAD$^+$] and [ATP]/[ADP][Pi]. An increase or decrease in intramitochondrial [NADH]/[NAD$^+$] at constant [ATP]/[ADP][Pi] causes a corresponding increase or decrease in respiratory rate. Thus *in vivo* regulation of the metabolic pathways generating intramitochondrial NADH has a major role in determining both the cellular energy level ([ATP]/[ADP][Pi]) and the respiratory rate (rate of ATP synthesis).

5.5. *The oxygen concentration dependence of the mitochondrial respiratory rate*

One of the most important unanswered biochemical and physiological questions is the oxygen dependence of cellular metabolism. This has been approached from two directions: measuring the dependence of the respiratory rate and other metabolic parameters on oxygen concentration and developing a model for the mechanism of cytochrome oxidase and using it to design optimal experimental tests for this dependence. The latter approach has proven very valuable because when the kinetics of the proposed mechanism for oxygen reduction by cytochrome oxidase [53] was examined, it predicted that the oxygen dependence should be much greater for cellular conditions of [NAD$^+$]/[NADH], [ATP]/[ADP][Pi] than those normally present for suspensions of isolated mitochondria [53]. Direct measurements of the oxygen concentration dependence of the reduction of cytochrome c, cellular [ATP]/[ADP][Pi] and intramitochondrial [NAD$^+$]/[NADH] in suspensions of intact cells support this prediction [53].

The rate of respiration and level of cytochrome c reduction in cell suspensions were measured using a spectrophotometer cuvette equipped with a stirrer and an oxygen electrode. This cuvette was placed in a dual wavelength spectrophotometer and cytochrome c reduction (measured at 550 nm minus 540 nm) was recorded simultaneously with the oxygen concentration in the cell suspension. Cellular ATP and ADP as well as the intramitochondrial [NAD$^+$]/[NADH] were measured in parallel experiments in which aliquots of the suspensions were rapidly ejected into cold deoxygenated perchloric acid solution to quench the enzymatic activities and then the metabolites assayed by standard techniques.

The data obtained for suspensions of cultured neuroblastoma C-1300 cells are summarized in Fig. 12. The respiratory rate of the cell suspension is essentially independent of oxygen tensions until the latter falls below 20 μM. A slow decrease occurs with further decrease in oxygen tension until a rapid decline takes place below 3 μM giving an apparent K_m^* for oxygen of less than 1.0 μM. Cytochrome c reduction is readily measurable when the oxygen tension falls from 100 μM to 30 μM and the extent of reduction increases with decreasing oxygen tension. The cellular energy supply expressed as [ATP]/[ADP] (intracellular Pi was assumed to be essentially constant as for technical reasons it can not be measured in this experiment) is measurably decreased from its high oxygen value of 10 when the oxygen tension decreases to 20 μM, but falls only to approximately 4 even in brief (30 s) periods of anaerobiosis. At this point glycolytic activity is sufficient to sustain temporarily the ATP level. The behavior pattern seen in Fig. 12 is not specific for cultured neuroblastoma cells as similar results have been obtained for suspensions of cultured kidney cells, cultured oligodendroglia, ascites tumor cells and *Tetrahymena pyriformis*.

---

* K_m, the MICHAELIS-MENTEN constant, is equal to the substrate concentration giving the half-maximal reaction velocity; in this case it is the oxygen tension at which the observed respiratory rate would be 50 % of the maximal rate.

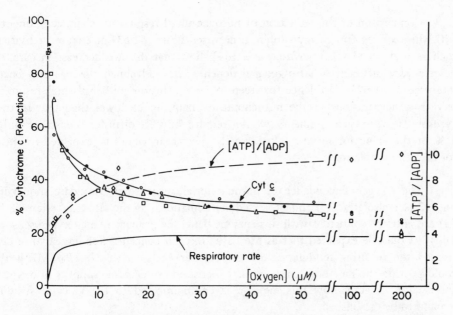

Fig. 12. — The oxygen tension dependence of the respiratory rate, cellular [ATP]/
[ADP][Pi] and cytochrome c reduction in suspensions of cultured neuroblastoma
cells. The cells were suspended at 25-30 mg wet weight/cm^3 in a HANK's medium.
Reduction of cytochrome c was measured at 550 nm minus 540 nm while [ATP]
and [ADP] were measured in samples which had been rapidly quenched and depro-
teinized in cold perchloric acid and then neutralized. The cytochrome c data from
4 separate experiments are plotted in order to indicate the reproducibility of the
data (taken Ref. 26)

The expression which describes the kinetic behavior of the proposed mechanism for oxygen reduction by cytochrome c oxidase (Refs. 52 and 53) fits accurately
the behavior of isolated mitochondria and cells at high oxygen tensions (Ref. 52).
When this expression is extended to incorporate the oxygen dependence (Ref. 53)
the kinetic behavior of the model is found to predict that K_m for oxygen should
markedly be dependent on both the [ATP]/[ADP][Pi] and the intramitochondrial
[NAD$^+$][NADH], becoming larger with increasing [ATP]/[ADP][Pi] and with increasing [NAD$^+$]/[NADH]. The final equation is entirely consistent with the observed oxygen dependence of the metabolic parameters of cell suspensions.

The values of the K_m of cytochrome c oxidase for oxygen reported in the
literature vary widely, quite possibly due to real variation in experimental conditions
[55-61]. This is expected if the K_m for oxygen is a function of the cellular metabolic
state.

Comparison of the reduction of mitochondrial respiratory chain components with the oxygenation of myoglobin in perfused heart [62-63] or enzyme activities such as urate oxidase in perfused liver [64] show that the two decrease in parallel despite very different *in vitro* oxygen dependencies. Although this has also been interpreted [61-62] as evidence for steep oxygen diffusion gradients which result in oxygen concentrations at the mitochondrion being much lower than that in the cytosol, such oxygen gradients would require oxygen diffusion constants only 1 % of the value for normal saline, much lower than would be expected to exist for the cell cytoplasm.

The oxygen dependence of mitochondrial cytochrome *c* oxidase shown in Fig. 12 (see also Ref. 55) and expressed mathematically in the rate expression (Ref. 53) are entirely consistent with this enzyme being the primary tissue oxygen sensor. Thus it would be expected that as the tissue oxygen concentration increased or decreased the resulting modulation of the cellular energy level ([ATP]/[ADP][Pi]) could provide the metabolic message that increased or decreased supplies of oxygen are needed. Exactly how this message is communicated to the vascular system remains to be established.

Summarizing it can be said that mitochondrial oxidative phosphorylation is of central importance to higher organisms because it is responsible for providing most of the metabolic energy (as ATP) which is required for their survival. Catabolism of organic substrates which reach the cell occurs in two phases. In the first, the substrates (sugars, amino acids and fatty acids) are degraded to 2-4 carbon units; in the second, (the tricarboxylic acid cycle) these smaller units are oxidized completely to CO_2. The oxidative steps in each phase involve enzymatic transfer of reducing equivalents from the catabolite to the intracellular pool of oxidation-reduction coenzymes, primarily NAD and flavins. The mitochondrial respiratory chain is concerned with reoxidation of these coenzymes, by using the reducing equivalents to reduce molecular oxygen to water. The NAD couple has an oxidation-reduction potential near -0.28 volts relative to a standard hydrogen electrode while the oxygen reaction is near 0.82 volts relative to a standard hydrogen electrode. The electron transfer process thus occurs across a potential span of 1.1 volts and is highly exergonic. A large fraction (70-80 %) of this energy is conserved by using it to drive the endergonic synthesis of ATP. The mitochondrial respiratory chain consists of several oxidation-reduction components and functions as an ordered sequence of reactions with energy conservation occurring at three discrete *sites*. These energy conservation sites are called sites I, II, and III, numbering from the more negative to the more positive end of the reaction sequence. In keeping with this observation three moles of ATP, one for each site, are synthesized for each two reducing equivalents transferred from NADH to molecular oxygen. The enzymes of oxidative phosphorylation are an integral part of the inner mitochondrial membrane and

ATP is synthesized on the inner side of this membrane (matrix space) and then translocated across the membrane (in exchange for ADP) to the cytoplasm where most of the ATP requiring reactions are located. (For general review see LEHNINGER [65], WAINIO [66], ERECINSKA and WILSON [67], SCHÄFER and KLINGEN-BERG [68], and VAN DAM and VAN GELDER [69]).

References

[1] P.L. DUTTON and D.F. WILSON, *Biochim. Biophys. Acta* **346**, 165 (1974).

[2] P.L. DUTTON, *Methods of Enzymol.* **54**, 411 (1978).

[3] D.F. WILSON, *An Approach to the Study of Electron Transport Systems in Methods in Membrane Biology*, E.D. Korn, (Editor) **10**, (1979) 181-222.

[4] D.F. WILSON and P.L. DUTTON, *Biochem. Biophys, Res. Commun.* **39**, 59 (1970).

[5] M. ERECINSKA, R. OSHINO, N. OSHINO and B. CHANCE, *Arch. Biochem. Biophys.* **157**, 431 (1973).

[6] P.L. DUTTON, *Biochim. Biophys. Acta* **226**, 63 (1971).

[7] D.F. WILSON, M. ERECINSKA, P.L. DUTTON and T. TSUDZKI, *Biochem. Biophys. Res. Commun.* **41**, 1273 (1970).

[8] J.S. Jr. LEIGH, D.F. WILSON, C.S. OWEN and T.E. KING, *Arch. Biochem. Biophys.* **100**, 476 (1974).

[9] D.F. WILSON, M. ERECINSKA and C.S. OWEN, *Arch. Biochem. Biophys.* **175**, 160 (1976).

[10] B.F. VAN GELDER and H. BEINERT, *Biochim. Biophys. Acta* **189**, 1 (1969).

[11] D.F. WILSON, J.G. LINDSAY and E.S. BROCKLEHURST, *Arch. Biochem. Biophys.* **151**, 180-187 (1972).

[12] M. ERECINSKA, D.F. WILSON and Y. MIYATA, *Arch. Biochem. Biophys.* **177**, 133 (1976).

[13] W.R. HEINEMAN, B.J. NORRIS and J.F. GOELZ, *Anal. Chem.* **47**, 79 (1975).

[14] D.F. WILSON and D. NELSEN, *Biochim. Biophys. Acta* **680**, 233 (1982).

[15] B.F. VAN GELDER, *Biochim. Biophys. Acta* **118**, 36 (1966).

[16] D.F. WILSON, M. ERECINSKA and P.L. DUTTON, *Ann. Rev. Biophys. Bioenerg.* **3**, 203 (1974).

[17] J.G. LINDSAY, C.S. OWEN and D.F. WILSON, *Arch. Biochem. Biophys.* **169**, 492 (1975).

[18] D.F. WILSON, C.S. OWEN and A. HOLIAN, *Arch. Biochem. Biophys.* **182**, 749 (1977).

[19] T.H. STEVENS, G.W. BRUDVIG, D.F. BOCIAN and S.I. CHAN, *Proc. Natl. Acad. Sci. USA* **76**, 3320 (1979).

[20] J.O. ALBEN, P.O. MOH, F.G. FIAMINGO and R.A. ALTSHULD, *Proc. Natl. Acad. Sci. USA* **78**, 234 (1981).

[21] L. POWERS, B. CHANCE, Y. CHING and P. ANGIOLILLO, *Biophys. J.* **33**, 95a (1981).

[22] D.F. WILSON, C.S. OWEN and M. ERECINSKA, *Arch. Biochem. Biophys.* **195**, 494 (1979).

[23] M. ERECINSKA and B. CHANCE, *Arch. Biochem. Biophys.* **151**, 304 (1972).

[24] Y. ORII and T.E. KING, *FEBS Letts.* **21**, 199 (1972).

[25] B. CHANCE, C. SARONIO and J.S. Jr. LEIGH, *Proc. Natl. Acad. Sci. USA* **72**, 1635 (1975).

[26] B. CHANCE, C. SARONIO and J.S. Jr. LEIGH, *J. Biol. Chem.* **250**, 9226 (1975).

[27] R.W. SHAW, R.E. HANSEN and H. BEINERT, *Biochim. Biophys. Acta* **548**, 386 (1979).

[28] T. OHNISHI in *Membrane Proteins in Energy Transduction*, R.A. Capaldi (Editor) Marcel Dekker, Inc. N.Y. (1979) pp. 1-87.

[29] P.L. DUTTON and D.F. WILSON, *Biochem. Biophys. Acta* **346**, 165 (1974).

[30] D.F. WILSON in *Membrane Structure and Function*, E.E. Bittar (Editor) John Wiley, N.Y. (1980) pp. 153-195.

[31] D.F. WILSON and M. ERECINSKA, *Mitochondria/Biomembranes*, Elsevier-North Holland, N.Y. (1972) pp. 119-132.

[32] D.F. WILSON, M. ERECINSKA, C.S. OWEN and L. MELA, *Dynamics of Energy Transducing Membranes*, L. Ernster, R.W. Estabrook, and E.C. Slater, (Editors), Elsevier N.Y. (1974) pp. 221-231.

[33] P.D. BOYER, B. CHANCE, L. ERNSTER, P. MITCHELL, R. RACKER and E.C. SLATER, *Ann. Rev. Biochem.* **46**, 955 (1977).

[34] G.D. GREVILLE, *A Scrutiny of Mitchell's Chemiosmotic Hypothesis of Respiratory Chain and Photosynthetic Phosphorylation*, D.R. Sanadi (Editor) *Current Topics in Bioenergetics* (1969) Vol. 3, Academic Press, N.Y. pp. 1-78.

[35] C.S. OWEN and D.F. WILSON, *Arch. Biochem. Biophys.* **161**, 581 (1974).

[36] A. HOLIAN, C.S. OWEN and D.F. WILSON, *Arch. Biochem. Biophys.* **181**, 164 (1977).

[37] M. ERECINSKA, R.L. VEECH and D.F. WILSON, *Arch. Biochem. Biophys.* **160**, 412 (1974).

[38] D.F. WILSON, M. STUBBS, R.L. VEECH, M. ERECINSKA and H.A. KREBS, *Biochem. J.* **140**, 57 (1974).

[39] D.F. WILSON, M. STUBBS, N. OSHINO and M. ERECINSKA, *Biochemistry*, **13**, 5305 (1974).

[40] M. ERECINSKA, D.F. WILSON and K. NISHIKI, *Am. J. Physiol., Cell Physiol.* 3(2) (1978) C82-C89.

[41] T. BENZINGER, C. KITZINGER, R. HEMS and K. BURTON, *Biochem. J.* **71**, 400 (1959).

[42] R. GUYNN and R.L. VEECH, *J. Biol. Chem.* **248**, 6966 (1973).

[43] D.F. WILSON, M. ERECINSKA, C. DROWN and I.A. SILVER, *Am. J. Physiol., Cell Physiol.* 2(3) (1977) C135-140.

[44] M. ERECINSKA, T. KULA and D.F. WILSON, *FEBS Letts.* **87**, 139 (1978).

[45] J.E. DAVIS and W.I.A. DAVIS VAN THIENEN, *Biochem. Biophys. Res. Commun.* **83**, 1260 (1978).

[46] M. KLINGENBERG, *J. Membr. Biol.* **56**, 97 (1980).

[47] E.C. SLATER, J. ROSING and A. MOL, *Biochim. Biophys. Acta* **292**, 534 (1973).

[48] R. VAN DER-MEER, T.P.M. AKERBOOM, A.K. GROEN and J.M. TAGER, *Eur. J. Biochem.* **84**, 421 (1978).

[49] M. ERECINSKA, M. STUBBS, Y. MIYATA, C.M. DITRE and D.F. WILSON, *Biochim. Biophys. Acta* **462**, 20 (1977).

[50] I. SUSSMAN, M. ERECINSKA and D.F. WILSON, *Biochim. Biophys. Acta* **591**, 209 (1980).

[51] H.A. KREBS and R.L. VEECH, *The Energy Level and Metabolic Control in Mitochondria*, S. Papa, J.M. Tager, E. Quagliariello and E.C. Slater, (Editors), Adriatica Editrice, Bari, Italy, (1969), pp. 329-382.

[52] D.F. WILSON, C.S. OWEN and A. HOLIAN, *Arch. Biochem. Biophys.* **182**, 749 (1977).

[53] D.F. WILSON, C.S. OWEN and M. ERECINSKA, *Arch. Biochem. Biophys.* **195**, 494 (1979).

[54] M. ERECINSKA and D.F. WILSON, *J. Membr. Biol.* in press (1982).

[55] D.F. WILSON, M. ERECINSKA, C. DROWN and I.A. SILVER, *Arch. Biochem. Biophys.* **195**, 485 (1979).

[56] N. OSHINO, T. SUGANO, R. OSHINO and B. CHANCE, *Biochim. Biophys. Acta* **368**, 298 (1974).

[57] A. BAENDER and M. KIESSE, *Arch. Exp. Pathol. Pharmakol.* **244**, 312 (1955).

[58] H. DEGN and H. WOHLRAB, *Biochim. Biophys. Acta* **245**, 347 (1971).

[59] I.S. LONGMUIR, *Biochem. J.* **65**, 378 (1957).

[60] L.C. PETERSON, P. NICHOLLS and H. DEGN, *Biochem. J.* **142**, 247 (1974).

[61] D.P. JONES and H.S. MASON, *J. Biol. Chem.* **253**, 4874 (1978).

[62] B. CHANCE, *Circ. Res.* **38**, 131 (1976).

[63] M. TAMURA, N. OSHINO, B. CHANCE and I.A. SILVER, *Arch. Biochem. Biophys.* **191**, 8 (1978).

[64] J. SIES, *Adv. Exp. Med. and Biol.* **94**, 561 (1978).

[65] A.L. LEHNINGER, *The Mitochondrion*, Benjamin, New York (1966).

[66] W.W. WAINIO, *The Mammalian Mitochondrial Respiratory Chain*, Academic Press, New York (1970).

[67] M. ERECINSKA and D.F. WILSON (Eds.), *Inhibitors of Mitochondrial Function. Section 107, International Encyclo. of Pharmacol and Therapeutics*,

Pergamon Press, New York (1981).

[68] G. SCHÄFER and M. KLINGENBERG (Eds.)., *Energy Conservation in Biological Membranes. 29th Colloquium der Gesellschäft Für Biologische Chemie.* Springer Verlag, Berlin, New York (1978).

[69] K. VAN DAM and B.F. VAN GELDER (Eds.), *Structure and Function of Energy Transducing Membranes.* Elsevier-North Holland Biomedical Press. Amsterdam (1977).

ELECTROCHEMICAL METHODOLOGIES IN BIOMEDICAL APPLICATIONS

J. RAGHAVENDRA RAO

Research Laboratories of SIEMENS A.G.
D-8520 Erlangen, Fed. Rep. Germany

Contents

1.   Introduction

Electrochemical systems implanted in living organisms can contribute to the various functions of biomedical research and equally to the field of diagnosis and therapy. The bioenergetics and energy requirements of living organisms have been treated thoroughly and impressively in other lectures of this course. Hence, it is evident that the metabolic equilibrium is maintained by a complex variety of biological oxidation and reduction reactions controlling the energy transfer to preserve the normal state of the living organism; what we generally denote as a healthy condition. Any imbalance and disturbance in this process would lead to metabolic and functional disorders mostly of clinical significance.

Temporary and local disorders can be counteracted by classical chemotherapy of medicines and the anatomical disease even by surgical invasion. However, if the disorder is chronic, affecting the physiology of body, it can only be encountered by a permanent and sustained clinical aid. Very often external means and methods are required for such a therapy. So, for example, some kind of a constant clinical device is indispensable in providing insulin for a body which by itself is unable to produce its own insulin requirement. A heart — otherwise anatomically sound — is bound to fail if natural impulses due to synergic interactions are not generated and propagated in it from one region to the other parts of muscular fibres. In such a case, an implanted pacemaker saves the life by constantly imparting artificial impulses to the heart in stimulating its functional activity. These examples show already the crucial need of methodologies as long-term life supporting or saving systems.

A special aspect of interest is to convert the biological energy into electrical energy. This bioelectrochemical conversion will serve as an inherent and internal power source to operate implantable systems. The procedures are more or less based on electrochemical principles, necessarily involving different types of electrodes. Therefore, in this lecture, some electrochemical methodologies, their principle, performance and scope as life supporting systems will be described.

2.   General remarks

Electrical energy is required to operate sensor electrodes for *in vivo* measurements, for biological growth control, for stimulating muscles and nerves, for monitoring and telemetric readings, and especially for stimulating the cardiac muscle. Efforts have even been made to generate more than 5 W of body power to drive a totally implantable artificial heart. It will perhaps be of interest to show in Fig. 1 the power requirement for different cardiac activities in the human body.

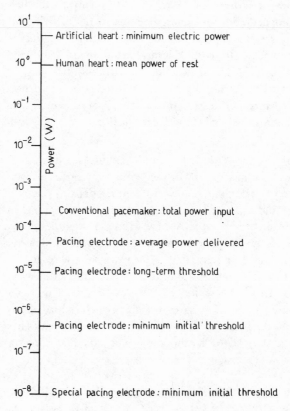

Fig. 1. — Power levels required for different cardiac activities in the human body

Apart from the power requirement and functional concepts, it is indeed of vital importance that the electrodes and other materials implanted to achieve the aimed purpose are essentially biocompatible. The electrode ability and activity, on one side, and its biocompatibility, on the other, are very difficult criteria to fulfil simultaneously and often compromises have to be made. Therefore, the question of biocompatibility stretches through all pertinent investigations like an unbroken thread with two basic aspects:

a)   the action of the body towards the foreign material (acceptance, tolerance or rejection) and

b)   the reaction of the foreign material inside the body (hostile and uncompatible reactions, chemical degradation or degeneration, loss of functional ability etc.).

Except for some empirical knowledge little is known systematically about these interactions so that one is almost left alone to make one's own practical experience.

One more disturbing factor in the optimal performance of the electrodes is given by the homeostatic parameters such as pH, temperature and chemical composition of blood and body fluids; leaving only a moderate scope to derive the full advantage of electrode characteristics.

In spite of these limitations and restrictions, the feasibility of some clinical aids and especially the role played by electrochemistry in biomedical research will be underlined and illustrated particularly from topics under pursuit in our laboratories.

3.   Energy sources and power generation

It has already been mentioned that an implantable device such as cardiac pacemaker requires a power source. Fig. 2 shows the basic design concept of implantable electrochemical energy sources.

Completely encapsulated primary cells have always been the preferred choice as a power source for cardiac pacemakers; because they are tightly sealed, without risk and any interaction with the body surroundings is thus eliminated. Initially, HgO-Zn cells were used with a clinical life of 2 years. But they have now been replaced by Li-I_2 cells containing a non-aqueous electrolyte. The high energy density coupled with the very low rate of self discharge is quite advantageous and the clinical life has thus been extended from 2 years to nearly 10 years. These primary cells contain the active mass stored within the electrodes.

In contrast, the body itself plays an active role in reaction systems where the active mass (reactant) could be derived partially or completely from the metabolism of biological systems. Biogalvanic and biofuel cells are two examples for this kind of electrochemical energy conversion. Both cell types contain a pair of electrodes, an anode (negative mass) and a cathode (positive mass), and operate in the body fluids as electrolytes.

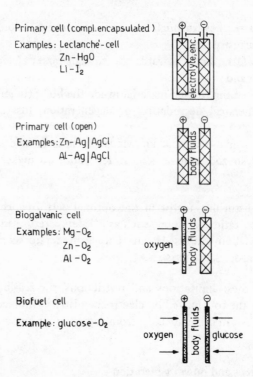

Primary cell (compl.encapsulated)

Examples: Leclanché-cell
        Zn – HgO
        Li – I$_2$

Primary cell (open)

Examples: Zn– Ag | AgCl
        Al– Ag | AgCl

Biogalvanic cell

Examples: Mg – O$_2$
        Zn – O$_2$
        Al – O$_2$

oxygen

Biofuel cell

Example: glucose – O$_2$

oxygen   glucose

Fig. 2. – Basic design concept of implantable electrochemical energy sources

In a true biogalvanic system, *i.e.* an open primary cell, both electrodes are sacrificial, *i.e.* they are used up during the course of electrochemical reactions. The life time is therefore determined by the amount of active mass of electrodes. Thus, in view of the cell size, the life period will be small.

In a metal oxygen type of a biogalvanic cell, only one electrode, *viz.* a metal anode is consumed (oxidative corrosion) while the catalyst cathode merely reduces the oxygen present in body fluids. The theoretical life of this cell is therefore determined only by the rate at which the anode corrodes. This type of a metal oxygen cell is also called a *hybrid cell* because one of its electrode, the anode, is of true galvanic nature and the other, the cathode, is a catalyst electrode of the fuel cell type.

In a biofuel cell both electrodes are catalytically active and the electrocatalysts convert reactants present in body fluids in order to generate electrical energy: the cathode reduces the oxygen and the anode oxidizes a fuel like glucose. The life of a biofuel cell is determined by the stability and continued efficiency of the catalysts and the transport of reactants to the electrode surfaces. If these factors are assured, the cell can operate indefinitely.

Biogalvanic cells are suitable for low power requirements (up to a few hundred milliwatts) whereas biofuel cells are also basically capable of generating higher powers exceeding 5 W. In the following, these two systems, *i.e.* biogalvanic metal oxygen and biofuel cells will be considered in more detail. However, regarding the biofuel cells, a brief remark is necessary at this place to avoid any misunderstanding.

The term *biofuel cell* used in this context refers to the *in vivo* electrochemical energy conversion (of intracorporeal substances) on fuel cell principle, *i.e.* direct conversion of chemical energy into electrical energy. Body oxygen and a *biological fuel* such as glucose are utilized and the fuel *cell* operates in biological media. Hence the term biofuel cell.

The term *biofuel cell* is also used elsewhere in a different context with a significant difference that in such a cell the reactions of one or of both electrodes are catalyzed by biological catalysts. The basic catalyst in all biochemical processes occurring in this kind of a biofuel cell is an enzyme or a group of interacting enzymes which perform molecular oxidation-reduction reactions. In this biofuel cell, by the action of microorganisms, a conversion of primary or secondary fuels to tertiary fuels (H_2, NH_3 or CH_4) occurs through biological metabolism. Tertiary fuels are then subjected to electrochemical oxidation by means of electrodes. The electrode reactions in this kind of cells are of different nature than the ones taking place in the biofuel cell under consideration (interface, phase boundary etc.). Therefore, biofuel cells based on microbial activities will not be considered here.

3.1. Biogalvanic cell

In deriving biogalvanic power, oxygen supplied by the arterial system is utilized as the oxidant (positive mass) whereas a metal is set as the negative mass. Only Mg, Zn, and Al or their alloys are suitable as anodes, and materials used as cathode catalysts include phthalocyanine, silver, platinum and activated carbon. Some data of the anode materials are represented in Table 1.

The table shows that aluminium is a very good choice as an anodic material. Although platinum has been mostly used as a cathodic material, silver but especially activated carbon is of greater advantage as an oxygen catalyst because of the high selectivity and economical availability.

The cell and its reactions are:

Table 1. — Some data of anodic materials for biogalvanic cells

Metal	Atomic weight	Valency	Density g cm⁻³ at 20°C	H_2-over-potential η at 25°C; 1 mA cm⁻² mV	Theoretical charge Q Ah g⁻¹	Theoretical charge Q Ah cm⁻³	Expected reactions in neutral medium (according to Pourbaix)	Oxide-electrode pot. $U = U_0 - 0.058\ pH$ V (at 25°C)	Theor. cell voltage (V) $U_{O_2} \mp U_{Me}$	Energy content Wh g⁻¹	Energy content Wh at 20°C cm⁻³	Corrosion in 5 years at 100 µW w.r.t. theor. energy content	Toxicity; + toxic, - non-toxic, o doubtful
Al	26.98	3	2.7	665	2.98	8.05	$2\,Al + 3\,H_2O \rightarrow Al_2O_3 + 6\,H^+ + 6\,e^-$	−1.35	2.78	8.26	22.3	0.53	—
Mg	24.32	2	1.74	700	2.21	3.86	$Mg + 2\,H_2O \rightarrow MgO + 2\,H^+ + 2\,e^-$	−1.86	3.09	6.82	11.87	0.64	—
Zn	65.38	2	7.14	720	0.82	5.86	$Zn + H_2O \rightarrow ZnO + 2\,H^+ + 2\,e^-$	−0.44	1.67	1.37	9.78	3.2	o
Cu	63.54	2	8.96	480	0.84	7.56	$Cu_2O + H_2O \rightarrow 2\,CuO + 2\,H^+ + 2\,e^-$	+0.67	0.56	0.24	2.12	7.87	+(?)
Fe	55.85	3	7.87	400	1.44	11.33	$3\,Fe + 4\,H_2O \rightarrow Fe_3O_4 + 8\,H^+ + 8\,e^-$						+(?)
							$2\,Fe_3O_4 + H_2O \rightarrow 3\,Fe_2O_3 + 2\,H^+ + 2\,e^-$	−0.047	1.28	1.23	9.65	3.57	+

(−) Al/NaCl-solution (porous catalyst electrode)/O_2 (+)

$$2\,Al + 6\,OH^- \longrightarrow 2\,Al(OH)_3 + 6\,e^- \quad \text{anodic partial reaction}$$

$$(3/2)\,O_2 + 3\,H_2O + 6\,e^- \longrightarrow 6\,OH^- \quad \text{cathodic partial reaction}$$

$$2\,Al + (3/2)\,O_2 + 3\,H_2O \longrightarrow 2\,Al(OH)_3 \quad \text{overall reaction}$$

$\Delta G^\circ = 1.56 \times 10^{-6}$ Joules/formula conversion; $U^\circ = 2.70$ V.

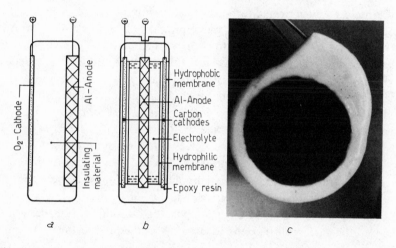

Fig. 3. — Biogalvanic Al-O_2-cells, a: open galvanic cell; cross-section, b: Chamber cell; cross-section; c: picture of an implantable model; weight 31 g, volume 12.4 cm^3

The experimentally designed biogalvanic cells are shown in Fig. 3. In the first type (a), the electrodes are cast in an epoxy resin but both electrodes (Ag-cathode and Al-anode) are free and open to have direct contact with body fluids. In vitro performance of such a *block cell* with an electrode area of 20 cm^2 yielded considerably low voltages because of polarization losses at both electrodes. A cell voltage of 0.6 to 0.8 V was obtained under the resistive load, corresponding to an average power load of 200 μW. However, under these conditions a life period of 10 years can be expected with a given weight of 10 g of pure aluminium (99.999 %). Animal experiments with this open cell, however, did not yield constant power as *in vitro* tests. Cell performance deteriorated and the power output fell considerably in the course of time. The formation of a less well-capillarized connective tissue around the cell limited oxygen transport to the electrode and thus restricted the power output. Since the electrodes were in direct contact with the surroundings, we assume that the insoluble oxidation-products at the electrode (formation of voluminous aluminium hydroxide) and a shift in pH at both electrodes might have stimulated uncontrolled growth of the connective tissue, and conversely, that biological substances might have interfered at the electrodes to deteriorate the cell performance.

Therefore, a better type of cell was designed what we call a chamber cell in which direct contact between the electrodes and their reaction products and body tissue is prevented. In this design an aluminium anode (b) is placed in the middle of the chamber and two porous carbon cathodes form the chamber walls on either side. Silver was replaced by activated carbon as a selective oxygen electrocatalyst. In a neutral medium, the electrode potential was approximately 200 mV more positive than with silver electrodes, thus leading to a higher cell voltage. The chamber is filled with isotonic physiological solution. The outer cathode surfaces are shielded with a body-compatible hydrophobic silastic membrane and the inner surfaces of the cathode facing the anode are protected with hydrophilic membranes (say, of polyvinyl alcohol). The outer hydrophobic membranes prevent any direct contact between the electrodes and body tissue or fluids, except for the oxygen supply, and the inner hydrophilic membranes maintain electrolytic contact between the electrodes while preventing the deposition of anodic oxidation products on the surface of the cathode. The covering hydrophobic layer must be very thin and also body compatible. Otherwise, connective tissue would grow too thickly and would strongly reduce the diffusion of oxygen. Thus, the cathodic limiting current density and thereby the maximum current load of the cell would no longer be sufficient.

Another advantage of this design is that several cells can be combined to form a battery and operated in series to produce higher voltage. Toxicity no longer restricts the choice of anode materials. The implantable cell (c) has the shape and size of a pocket watch; about 12.5 cm^3 in volume and 31 g in weight. The amount of aluminium is 5 g, sufficient to provide a service life of 10 years. Fig. 4 shows the *in vitro* characteristics of a chamber cell. The voltage and power density curves as a function of the current density show that maximum power is derived at 55 μA/cm^2; the voltage corresponding to 450 mV. After the successful *in vitro* testing, the chamber cells were implanted in animals. The cells were connected to different resistive loads and to a telemetric unit which could transmit even very low voltages. The cells were placed between the oblique external and internal abdominal muscles of dogs. Fig. 5 shows the *in vivo* performance of the biogalvanic cells.

A limiting current, determined by the supply of oxygen through the layers at the cathode, is established after about 300 days; with a load resistance of 2.7 kΩ and 4.7 kΩ respectively. The value lies around 70 μA, corresponding to a lim-

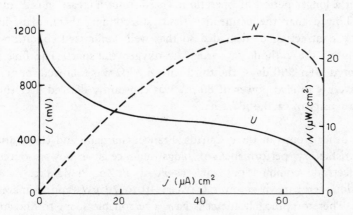

Fig. 4. — *In vitro* measurement with a biogalvanic chamber cell, cell voltage *U* (left ordinate) and power density *N* (right ordinate) depending on current density *j*. Al-anode, porous carbon cathode containing 5 wt % of Pt-Ru, physiological NaCl solution, 38 °C, p_{O_2} = 1 bar.

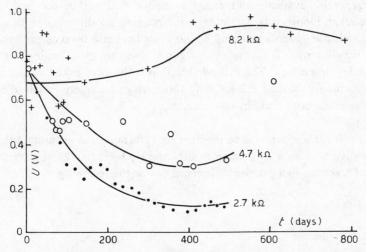

Fig. 5. — *In vivo* performance of biogalvanic chamber cells. Load resistances 2.7 kΩ, 4.7 kΩ and 8.2 kΩ respectively

iting current density of approximately 3 μA/cm^2. The load is only below the limiting current range at a resistance of 8.2 kΩ, the voltage then remains above 0.75 V even after a longer period of operation. The distinct increase in cell voltage after 300 days showed that the connective tissue surrounding the implant had either decreased by contraction or degraded so that well-capillarized condition persisted more closely to the cathode favouring the oxygen transport. The first two cells were explanted after 500 days, the third with 8.2 kΩ was followed for nearly three years. An average *in vivo* power of 80 μW was constantly derived over this period, *i.e.* a power density of ca. 3.4 μW/cm^2.

Even other groups in USA, Canada, France, Germany and Great Britain have reported satisfactory performance of biogalvanic cells as power sources. Using different electrode combinations such as Pt-Al, Pt-Zn, Pt-Mg etc. they derived an average long-term in vivo power between 40 to 70 μW (approximately 1.5 to 3 μW/cm^2). Therefore, biogalvanic cells have a reasonable scope to find application for low power and low risk implantable units.

3.2. Biofuel cells

The principle of a biofuel cell has already been outlined in Fig. 2. Both electrodes contain an electrocatalyst as an essential material which accelerates the desired electrode reaction; the fuel (glucose) anode and an oxygen cathode. In theory, all oxidizable substances present within the body and that readily exchange electrons with a catalyst electrode could be used as fuels. The oxidation of glucose, the most generally available and easily accessible fuel within the body; nearly 1 g/dm^3 (liter) in blood or body fluids, is the primary anodic reaction in a biofuel cell. Other fuels like glucosamine and amino acids have also been considered. These substances may contribute to the energy conversion but their limited availability makes them less important. The cathode reaction is the same as in biogalvanic cells. The reactants for the biofuel cell are only those which are freely dissolved in blood and body fluids and not bound in any manner.

Glucose is readily oxidized to the next step gluconic acid releasing 2 electrons. But quantitative oxidation of glucose to CO_2 and H_2O sets free 24 electrons. Therefore, the cell reactions in a glucose biofuel cell can be the following:

I $C_6H_{12}O_6 + 2\,OH^- \longrightarrow C_6H_{12}O_7 + H_2O + 2\,e^-$ anodic partial reaction

  $1/2\,O_2 + H_2O + 2\,e^- \longrightarrow 2\,OH^-$ cathodic partial reaction

---

  $C_6H_{12}O_6 + 1/2\,O_2 \longrightarrow C_6H_{12}O_7$ overall reaction

  $\Delta G^\circ = -\,2.51 \times 10^5$ J/mol; $U^\circ = 1.08$ V

or

II $C_6H_{12}O_6 + 24\,OH^- \longrightarrow 6\,CO_2 + 18\,H_2O + 24\,e^-$ anodic partial reaction

  $6\,O_2 + 12\,H_2O + 24\,e^- \longrightarrow 24\,OH^-$ cathodic partial reaction

---

  $C_6H_{12}O_6 + 6\,O_2 \longrightarrow 6\,CO_2 + 6\,H_2O$ overall reaction

  $\Delta G^\circ = -\,2.870 \times 10^6$ J/mol; $U^\circ = 1.24$ V.

The complete oxidation of glucose leads to an energy content of 2.87 MJ (or 686 kcal), *i.e.* 2.87/180 = 15.94 kJ (approximately 3.8 kcal) per gram of glucose.

Comparison of these two cell reactions (I and II) shows that at least 10 times more energy can be derived by oxidizing a glucose molecule quantitatively, although the theoretical cell voltage in both cases remains almost the same. The first cell reaction (I) is sufficient enough for low power requirements. But if the second reaction (II), *i.e.* conversion of 24 electrons per glucose molecule could be realized, a very energy rich fuel would be available and such an electrochemical reaction would widen the scope of biofuel cells to derive power in the range of 5 W required to drive a totally implantable artificial heart.

Now, before describing the design and performance of biofuel cells, it is worthwhile to consider the electrochemical conversion of glucose in generating, say, at a practical cell voltage of 0.5 V:

a) low power of 100 μW involving a 2-electron transfer, and

b) high power of 5 W involving a 24-electron transfer.

The amount of glucose required for these purposes can be given by FARADAY's law;

$$w = \frac{M\,I\,t}{n\,\text{F}}$$

where w is the amount of glucose to be converted, M its molecular weight (180), I = the current, t = the time (24 h or 1 d), n = the number of electrons transferred and F = FARADAY's constant; 26.8 Ah/mol or 96500 As/mol. Accordingly, the amount of glucose required will be;

for a) 16.12 mg/day or 0.187 μg/s and

for b) 67.16 g/day or 0.78 mg/s: this amount is present in less than 1 cm^3 of blood.

The blood circulation rate is 250 cm^3/min in heart, 1200 cm^3/min in skeletal muscle, 1100 cm^3/min in kidney, 750 cm^3/min in brain, 500 cm^3/min in skin and 600 cm^3/min in other organs. The regulative time for glucose level in the body is approximately 7 min. Similar estimations can also be made for oxygen conversion at 5 W of power. The rate of oxygen consumption will be roughly 5×10^{-5} mol/s. If the venous blood concentration of 3×10^{-3} mol/dm^3 is considered, 15 to 17 cm^3 of blood can formally supply this quantity. But since it corresponds to a pure oxygen content of 1.1 cm^3, approximately one-fifth of the blood circulation has to be supplied for this consumption. The temperature rise in blood due to heat losses at a 20 % conversion efficiency will range between 0.02 and 0.25 °C depending on the blood circulation rate (sp. heat of whole blood is 0.87 cal deg^{-1} g^{-1}). These considerations show the basic possibility of deriving a power of 5 W or even more in the body.

Any attempt to realize a biofuel cell immediately encounters a fundamental difficulty: both reactants, oxygen and glucose, are simultaneously present in blood and body fluids. The electrodes immersed in the common solution must therefore respond specifically to only one partner. In other words, selective electrocatalysts should be employed for the anode and the cathode respectively. There has been an intensive search for such materials. Success can be reported for oxygen electrodes only. Silver and especially activated carbon, for example, are active electrocatalysts for the cathode without unduly accelerating the anodic reaction. However, an equally specific and active anode catalyst is not available for glucose. Therefore, it is necessary to introduce an additional selecting component at the electrode in designing the fuel cell. The schematic design of such a cell is shown in Fig. 6.

The problem of selectivity can be solved according to this design by a special arrangement of electrodes. A non-selective anode like platinum-black is sandwiched between two thin, porous and selective oxygen cathodes. The electrodes, 250 μm in thickness, are separated by hydrophilic membranes permeable to oxygen, glucose and electrolytes. The outer electrode surfaces are also covered with thin hydrophilic membranes to prevent direct contact with body tissue or fluids. The outer membranes should be permeable to small molecules but not to larger ones like proteins.

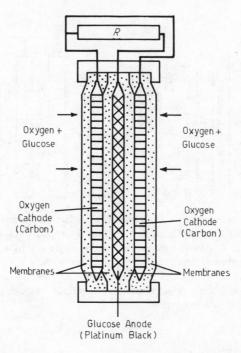

Fig. 6. — Diagram of electrode arrangement in a glucose oxygen biofuel cell

When the cell is in operation, the reactant mixture present in the electrolyte must be transported to the oxygen-consuming cathode before glucose can diffuse to the non-selective anode to undergo reaction without the interference of oxygen. The separation demands that oxygen be present in the mixture in much lower equivalent concentrations than glucose. This condition — as we have seen earlier — is completely fulfilled in blood and body fluids. The interfering limiting current of oxygen reduction is only 0.02 mA/cm^2 (at the usage of 12 μm thick membranes). Thus, it is smaller than the glucose oxidation current by an order of magnitude, even if only two electrons are transferred from the glucose molecules. Such a single cell can be constructed very thin and a number of these cells can be stacked in series to derive higher power. The design concepts of this cell battery are shown in Fig. 7.

The first one on the left was designed by GINER and coworkers in USA. Teflon-bonded platinum electrode is the cathode and is separated from the blood by a hydrophobic oxygen-selective silastic membrane. The hydrophobic membrane is structured in such a manner that blood channels are formed when single cells are stacked upon each other. The anode is a pure platinum or a noble metal alloy like Pt-Ru. It is covered with a hydrophilic cuprophane membrane permitting the trans-

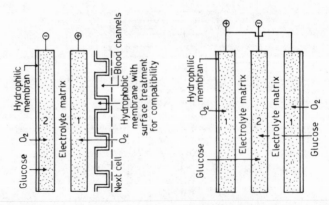

Fig. 7. (Left): − Cross-section of an individual biofuel cell, perpendicular to blood-flow, from GINER and HOLLECK. 1. porous hydrophobic O_2-cathode (catalyst Pt or Au-alloy); 2. porous hydrophilic glucose anode (catalyst Pt or alloys like Pt-Ru). Blood channel-depth ca. 100 μm, cell thickness ca. 250 μm. (Right): Cross-section of a biofuel cell with shielding cathodes, 1. porous hydrophilic O_2-cathode with selective electrocatalysts; 2. porous hydrophilic glucose anode with a non-selective electrocatalyst.

port of oxygen and glucose. The electrodes are held apart by means of an anion-exchanging membrane (electrolyte matrix). Though this is a practical design, a partial electrochemical short-circuit is unavoidable in this cell. This difficulty is not directly involved in the design shown on the right side. GINER and coworkers have also made the feasibility studies of their concept. Table 2 shows their conceptual data.

The practical performance of a biofuel cell is strongly dependent on the electrochemical oxidation of glucose. Apparently, the electrochemical oxidation of such a complicated molecule as glucose, which can basically release 24 electrons while undergoing various oxidation steps, is a complex process. The tentative reaction paths with some possible oxidation products are shown in Fig. 8.

Though an imaginary scheme, it emphasizes the importance of electrocatalytic activity of the electrodes. Platinized platinum catalyzes glucose oxidation to gluconic acid only and it is not stronger enough for further oxidation. As a result the oxidation product gluconic acid remains adsorbed on the catalyst surface causing electrode poisoning. For this reason, the performance of biofuel cells deteriorates rapidly. Thus, powerful catalysts are necessary for glucose oxidation and especially for quantitative conversion if higher power is desired. Therefore, electrocatalysts,

Table 2. — Power requirement and conceptual data for an implantable biofuel cell to power an artificial heart (GINER and HOLLECK)

Power requirement of human heart:
    mean power at rest   < 1 W
    mean power at moderate work   3.8 W
    instantaneous peak power (systole)
    at moderate work   10 W
Blood characteristic:
    O_2 concentration in plasma 1.35×10^{-4} mol/dm^3
    O_2 concentration in whole blood 8.8×10^{-3} mol/dm^3
    glucose (free) concentration 4.5×10^{-3} mol/dm^3
    Cl^- concentration $0.1 - 0.11$ mol/dm^3
    HCO_3^- concentration $0.024 - 0.030$ mol/dm^3
    PO_4^{3-} concentration $1.6 - 2.7 \times 10^{-3}$ mol/dm^3
    electrical conductivity (25 ° C) $0.012 \; \Omega^{-1}$ cm^{-1}

Biofuel Cell:
    Power output   5 W
    (hybrid configuration with NiCd battery
    for peak power)
    Volume   500 cm^3
    200 cells, area of each electrode 15×10 cm
    thickness of each cell   250 μm
    blood channel depth   100 μm
    electrolyte between electrodes   10 μm
    O_2 current, limited by diffusion (25 μm
    dimethylsilicone membrane, Δp_{O_2} = 0.13 bar)   0.8 mA/cm^2
    glucose current, limited by diffusion (12 μm
    cuprophane membrane)   0.2 mA/cm^2 ($n = 2$)
                                                                  2.4 mA/cm^2 ($n = 24$)
    cell voltage   0.5 V
    current density   1.0 mA/cm^2

particularly for glucose anode (in neutral media), are a very important criterion for the development of biofuel cells.

    The search for more powerful and active glucose catalysts showed that alloys of noble metals and ferrous metals provide reasonably well suited anode materials for the electrocatalytic oxidation of glucose. These metals such as Pt on one side

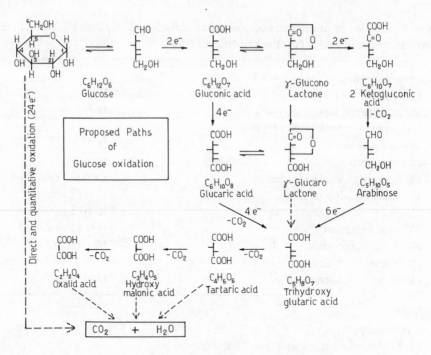

Fig. 8. — Imaginary paths of glucose oxidation

and Ni or F on the other, form homogeneous alloys in a broad range of composition which do not disintegrate in different phases and which can be rolled down to very thin, elastic but stable foils (50 μm). The catalytic activity of these alloys is even improved by doping them with transition metals such as W, Ti and Ta. A high specific surface area (up to 40 m^2/g; for Pt = 20 m^2/g) is finally obtained by extracting the ferrous metal from the foil (activation by anodic oxidation). The scanning electron-micrograph in Fig. 9 shows the catalytic structure obtained for a Pt:Fe alloy (atomic ratio 1 : 3).

The thickness of the active catalyst layer in 55 μm. The micropores responsible for the activity have not been resolved even at this magnification. Their radii are approximately 2 nm. Only the secondary macropores are visible which here are oriented perpendicularly to the surface. That is the good reason for the adherence of the active layer on the metallic support. These active layers are in fact surprisingly strong and are hardly damaged by scraping. Similarly, active catalysts were also obtained with either Pt-Ni (1:6) or their doped alloys. Fig. 10 shows the catalytic activity of some of these catalysts towards glucose oxidation in steady-state.

Fig. 9. — Scanning electron-micrograph of Pt-Fe electrode; Pt:Fe = 1:3; potentio-
statically activated in 1 M H_2SO_4; thickness of the active layer: 55 μm

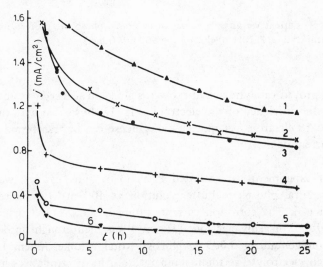

Fig. 10. — Current density-time curves for glucose oxidation in deaerated phosphate
buffer solution. Applied potential 0.4 V (r.h.e.); 0.1 mol/dm^3 glucose. 1: Pt W-Ni
(1:4); 2: Pt-Ni (1:6); 3: Pt/Ta-Ni (1:4); 4: Pt-Ni (1:4); 5: conventional Pt-electrode
and 6: conventional fuel cell electrode (AA 1)

The initial high values above 1 mA/cm² (projected for artificial heart) cannot be maintained in long-term operation. Nevertheless, the current densities referred to a comparable amount of Pt are higher at least by a factor of 8 to 10 than with conventional Pt-black electrodes of fuel cells. Fig. 11 even demonstrates the activity of the new catalyst Pt-Ni (1:6) towards the oxidation-products of glucose, such as gluconic acid.

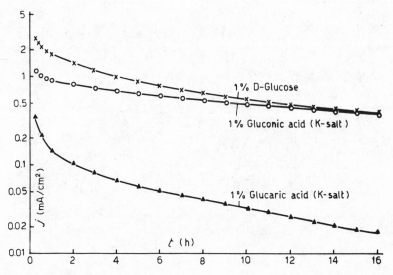

Fig. 11. — Current *versus* time curves in deaerated phosphate buffer solution at room temperature; Pt-Ni(1:6)-electrode at 400 mV (r.h.e.)

The improved catalytic activity towards gluconic acid seems to be due to high electron transfer efficiency of the electrode. Fig. 12 shows that with these catalysts at least 16 to 18 electrons per molecule of glucose can be transferred though with decreasing current density.

In this experiment, an activated Pt-Ni (1:6) alloy catalyst was previously tested in a neutral phosphate buffer solution at 400 mV [versus real hydrogen electrode (r.h.e.)]. The corrosion current dropped below 1 μA/cm² and was not measurable after 2 h. This electrode (15 cm²) was then brought into a phosphate buffer solution containing 0.005 M glucose (total volume 50 cm³). Argon was passed through electrolyte at room temperature. The electrode was maintained at 400 mV (versus r.h.e.) and the current flowing was recorded and integrated. The initial current of 1.6 mA/cm² decreased to 250 μA/cm² after 100 h; to 100 μA/cm² after 500 h and to 20 μA/cm² after 800 h. The experimentally derived charge was

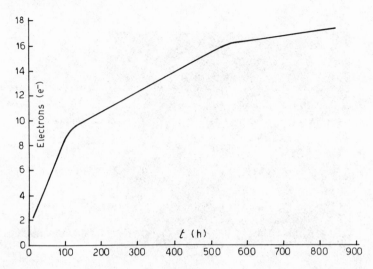

Fig. 12. — Oxidation of glucose. Electron-transfer per molecule of glucose as a function of time (estimated mean value of n). Electrolyte: phosphate buffer + 0.1 % glucose saturated with Ar at RT; potential applied: $U = 400$ mV (r.h.e.)

then compared with the ideal charge that could be obtained. This enables us to express the mean number of electrons transferred per glucose molecule in the course of the experiment. The n value thus obtained was 17.5. This, however, does not imply that every molecule of glucose released 17.5 electrons. Some molecules might have been oxidized to CO_2 whilst others were only partially oxidized.

Inspite of these developments, power generation up to 5 W is not realizable at present because of the lack of a catalyst system which can oxidize glucose quantitatively to CO_2 rapidly, $i.e.$ with a higher current density ($\geqslant 1$ mA/cm^2). Therefore, efforts were more diverted to develop biofuel cells for lower powers such as required by cardiac pacemakers.

Fig. 13 shows the section of a practical biofuel cell according to the principle already described. $In\ vitro$ performance of this cell is shown in the following Fig. 14. The result of a long-term test under varied partial pressures of oxygen, glucose concentration in different electrolytes is recorded in this figure. The voltage drop after 200 days might be associated with the separation of spacers from the electrodes and with beginning of formation of diffusion barriers. Also the oxygen electrodes have suffered some catalyst losses after this period. Microbes have repeatedly grown during the test period, though not damaging the cell.

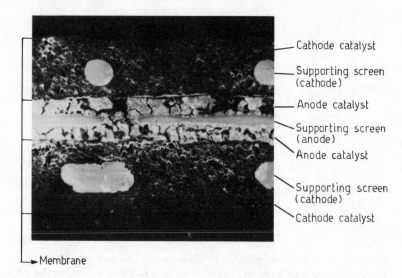

Fig. 13. – Scanning electron micrograph of a section through a biofuel cell (f = 60)

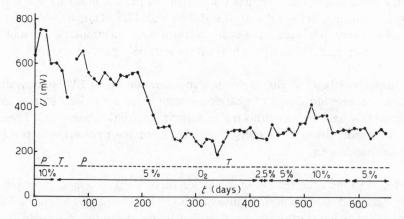

Fig. 14. – *In vitro* test with a biofuel cell. Electrode diameter 4 cm, load resistance 10 kΩ, 37 °C, P: phosphate buffer solution with 2 % glucose; T: Tyrode solution with 0.1 % glucose; O_2 content of the gas mixture ($N_2 + O_2$) bubbling through the solution has been varied according to the indication in the figure

Later, a cell constructed with greater precautions performed better, as shown in Fig. 15. The cell was operated with alternating load resistance and at different percentages of oxygen. The results showed that a sustained power of 60 μW (at 150 μA) or 5 μW/cm² (12,5 μA/cm²) could be derived over this period.

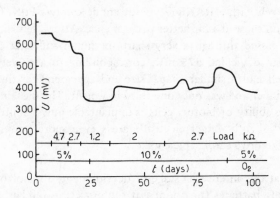

Fig. 15. — Long-term test of a biofuel cell in modified RINGER (Tyrode) solution containing 0.1 % glucose at 37 °C

After this successful *in vitro* test, biofuel cells were implanted in animals to test their *in vivo* performance. Two different biofuel cells were implanted along with telemetric units in two separate dogs in a similar manner as the biogalvanic cells and their cell voltage was recorded. Fig. 16 shows the results.

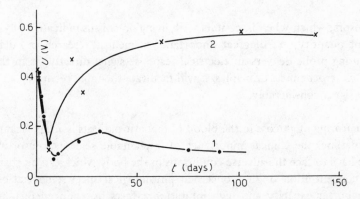

Fig. 16. — *In vivo* performance of biofuel cells implanted in dogs; 1: anode Pt-black 20 cm² in area, cathode activated carbon, 25 cm² in area, external load resistance 10 kΩ; 2: mechanically stabilized type, anode Pt-Ni foil with RANEY-layer, 10 cm², cathode activated carbon, 12.5 cm², external load resistance 8.2 kΩ

The performance of the cell containing Pt-black anode deteriorated rapidly to 100 mV. As described in details earlier, the reason for this poor performance lies in the poisoning of Pt-black surface with the oxidation product of glucose, *i.e.* due to a weaker catalytic activity of Pt-black. Moreover, the cell also suffered mechanical damages after 100 days. So, it had to be explanted.

The other cell with a RANEY-type of anode catalyst Pt-Ni (1:6) and a more stable construction showed a far better performance. After a few days (11 - 14 days) the cell voltage passed through a very sharp minimum and after 150 days of continuous performance yielded 575 mV, corresponding to a power of 40 μW or 4 μW/cm^2. This cell had a similar shape, size and appearence as the biogalvanic cell but was much thinner and was only one-third in weight. This sustained *in vivo* power demonstrates the ability of biofuel cells as implantable power sources. With a better modification and slightly larger electrode areas even more power can be derived than is necessary to drive cardiac pacemakers.

However, at this interesting stage of development, a major breakthrough was achieved with Li-I$_2$-batteries (mentioned at the outset) as implantable, reliable and uncomplicated power sources for cardiac pacemakers. Thus, biofuel cells seem to have lost their importance at present. But such efforts are seldom futile and often lead to make a virtue of necessity. The next section reveals particularly the extended scope of the biofuel cell principle: from a power generating cell to an indicator cell (glucose sensor). Moreover, the generation of biological power is still of actual interest, *e.g.* for implantable devices in controlled drug release.

4.   Sensor electrodes

Sensors, which when in contact with living organisms indicate a physiologically relevant quantity, are of great importance in medical research and diagnosis. If the measuring probe delivers an electrical response signal directly, as in the sensors based on electrochemical principles, it will facilitate the rapid recording and processing of the signal considerably.

Monitoring of glucose in the blood of diabetic patients is a significant example which underlines the clinical importance of implantable selective sensor electrodes. Again we have to face the adverse conditions in the body which can be characterized by: low concentration of the concerned physiological species, neutral electrolyte with low buffer capacity and high inhibition effects by endogeneous substances.

Until now, use has been made of methods which measure an electrode or a membrane potential without current, as for instance with ion selective electrodes, and methods which record a reduction or oxidation current under defined conditions. In the latter case, the current is generally limited by the diffusion of an electroactive substance through a layer permeable to this substance. Examples of electrochemical sensors along with their measuring principles are listed in Table 3.

Table 3. — Electrochemical sensors for biomedical application

Physiological substance to be determined	Measuring principle
O_2	polarographic measurement of diffusion current
CO_2	over pH-changes at the glass electrode
H^+ (pH)	glass electrode
$Na^+, K^+, Ca^{2+}, Cl^-, F^-$	ion selective electrodes
lactate	oxidation with $[Fe(CN)_6]^{3-}$ in the presence of cytochrome b_2; reaction product $[Fe(CN)_6]^{4-}$ is oxidized anodically
glucose	enzyme membrane with H_2O_2-anode
cholesterol (corresponding to polysaccharide)	enzymatic conversion with cholesterol oxidase to cholestenon and H_2O_2 which diffuses through the membrane to the cathode
urea	after the reaction in presence of urease, NH_4^+ ions measured with ion sensitive electrodes

The use of sensors to monitor blood gases, electrolytes and the metabolic substrates (*e.g.* lactate and pyruvate) is of great chemical interest. These parameters are significantly influenced by anesthesia and surgery. Continuous monitoring will provide information for rapid and rational decision making to improve patient care. Chemical monitoring will also be a great asset to patients with kidney failure. In hemodialysis and hemofiltration, the role of the molecules such as urea and creatinine is critical. Monitoring will help to optimize the solute removal rate from the patient. Chemical sensors also offer significant application in obstetrics, *e.g.* in the continuous monitoring of fetal biochemical variables during pregnancy. Sensors of fetal blood gases afford the opportunity of early detection of fetal distress and further increase our ability to study and monitor the fetus during pregnancy.

The scope of precise monitoring is significantly improved today by the introduction and application of solid state chemical sensors. These are in principle similar to ion-selective electrodes. A chemically sensitive field effect transistor

(CHEMFET) or ion-sensitive field effect transistor (ISFET) is obtained by replacing the conventional gate metal electrode of an IGFET (Insulated gate field effect transistor) with a chemically selective layer. Changes in potential (charge) at the interface of this layer with solution (or gas) are then reflected in the changes of the transistor drain current. Because of its sensitivity to ionic activities in solution coupled with its inherent property of extremely high input impedance, the CHEMFET (or ISFET) offers exceedingly good promise for monitoring many problems which limit the clinical applications of conventional chemical electrodes.

Among the physiological species listed in Table 3, glucose will be considered once again in detail because of its critical importance in *diabetes mellitus*. *Diabetes mellitus* is a disease due to metabolic disturbance and is characterized by the inability to keep the glucose level in the blood within the desired limits. It is based on an irregular or insufficient release of insulin by the beta cells causing an excessive sugar level in the diseased person. Harmful consequences of such a chronic disorder are accelerated arteriosclerosis and organ defects.

For some time, efforts have been under progress to compensate this hormonal malfunction by an open or closed-loop device, for supply of insulin. Such a system is called artificial pancreas or artificial beta cell. The closed-loop system proposed for total implantation consists of a glucose sensing unit, an insulin depot, a dosage unit, an energy source and an electronic device. An implantable glucose sensor is essentially required for this closed-loop system. Therefore, the feasibility of glucose sensor will be considered.

4.1. Glucose sensor

If the glucose level in the blood circulation of the patient is continuously monitored the clinically required adjustment of insulin will be facilitated. The monitoring of glucose can be done by two electrochemical methods:
i)   by enzyme sensor and
ii)   by electrocatalytic glucose sensor.

4.1.1. Enzyme sensor — The enzyme sensor is based on the enzymatic oxidation of glucose to gluconic acid in the presence of glucose oxidase according to the equation:

$$C_6H_{12}O_6 + H_2O + O_2 \xrightarrow{\text{glucose oxidase}} C_6H_{12}O_7 + H_2O_2$$

The glucose oxidase is immobilized in a membrane which covers a platinum electrode. This electrode measures either the consumption of oxygen during the reaction or the formation of hydrogen peroxide by eletrochemical means. Although glucose oxidase is a very specific enzyme catalyst for glucose oxidation, its suitability in long-term implantation is doubtful because of the gradual inactivation of the enzyme under body conditions. Moreover, its long-range stability under these conditions is also questionable. The method is an indirect one to detect, measure or monitor glucose concentration.

4.1.2. Electrocatalytic glucose sensor — The electrocatalytic sensor is a direct method to trace or estimate glucose and it is a product of the preceding intensive research work on biofuel cells. With a suitable design, the performance of a glucose-oxygen cell can also be easily made to depend on the glucose concentration in the surrounding medium. However, the biofuel cell is not directly applicable as a useful glucose sensor. The strong deactivation of the anode catalyst under the working conditions necessitates the use of voluminous quantities of catalyst. Thus, the processes of reaction and diffusion would be retarded and limited, leading to an intolerable time constant. The sensor would respond to a concentration change with considerable delay.

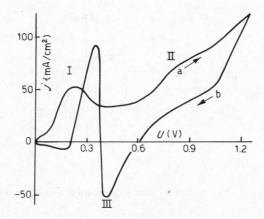

Fig. 17. — Potentiodynamic glucose oxidation with a Pt-electrode in phosphate buffer solution at pH 7, 25 °C, glucose concentration 0.2 %, sweep-rate 1 V/min

Fig. 17 shows the curve of potentiodynamic oxidation of glucose with a Pt-black electrode in phosphate buffer solution at 25 °C. The current *versus* voltage curve shows two regions at which the anodic current is determined by the glucose concentration: I at 200 mV, II at 500 to 800 mV (*versus* r.h.e.). During the backward sweep to negative electrode potentials (curve b), the potential range I is characterized by a steep current rise in anodic direction. The current rise sets in when the oxygen coverage at the Pt-surface is just eliminated by cathodic reduction (peak III). This experiment shows how the course of oxidation is essentially determined by the pre-treatment and the state of the electrode surface; an observation made with dissolved carbon-containing reactants, *e.g.* alchols. The experiment also indicates that the initial, high rate of reaction can be maintained by frequently rejuvenating the electrode, *i.e.* by repeated surface oxidation. A low catalyst-loading will then be adequate and the time constant of the sensor will no longer be determined by the electrode.

The oxidation currents of the curve are kinetically limited; they do not exhibit linear dependence on the concentration, are difficult to reproduce and therefore are not suitable for undisturbed analytical evaluation. If a diffusion barrier for glucose is created in front of the working electrode, the transport can be restricted to the extent that the kinetics would become no longer determining. This idea can be realized by covering the active electrode surface with a hydrophilic membrane of an appropriate thickness. The design of such membrane-covered glucose sensor is shown in Fig. 18.

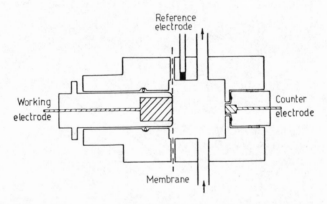

Fig. 18. — Glucose sensor in an experimental flow through cell

It represents a flow-through cell, consisting of a working, reference and a counter electrode. In an implantable sensor all these three electrodes are integrated into a small unit with the ultimate aim of reducing the whole unit to a diameter of 1 mm so that it could perforate the skin or a vein, like an injection needle. This pin, the tip of which is platinized, serves as the working electrode for the measurement. In addition, the tip of the pin is also covered with a diffusion limiting membrane. In the actual measurement, a potential programme is imposed upon this pin-electrode by a potentiostat and a reference electrode and the programme is determined by the current flowing at the counter electrode. Fig. 19 shows the circuit-diagram of the setup to measure glucose with two controlled potential sources at working and rejuvenating potentials respectively.

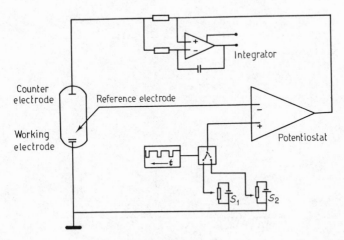

Fig. 19. — Basic circuit diagram of the setup for measurement; S_1 and S_2, two control voltage sources; t, time-programme control

In this so called potential-jump method, the electrode potential was period-ically reversed between the working and the rejuvenating potentials. The working and the rejuvenating potentials of the working electrode are preset or fixed by means of two (control) voltage sources (S_1 and S_2 respectively) and reversed with the help of a time programme transmitter. The potentiostat regulates the electrode potential against the reference electrode in accordance with the preset control voltage. The current flowing through the counter electrode and the integral of this current are registered by a recorder. The potential programme is schematically shown in Fig. 20.

The current *versus* time curve is depicted above by the dashed lines while the full-line curve yields the integral during the period of measurement. After reverting to working potential, the integrator is switched with delay in order to separate the capacitive currents which emerge due to charge reversal of the electro-chemical double-layer and from the reduction of platinum oxide. A constant, overlapped charge exchange which, for instance, results from the reduction of oxygen in the electrolyte is compensated.

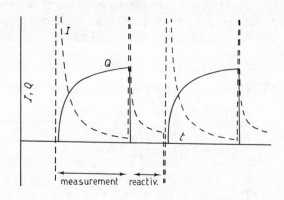

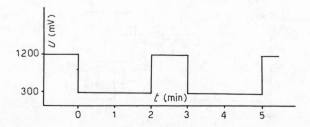

Fig. 20. – Potential-programme and the recorded current (*I*) as well as the charge (*Q*) with time at a glucose-catalyst electrode

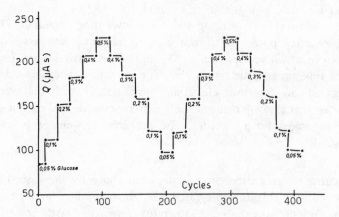

Fig. 21. – Course of response by the potential-jump method at varying concentrations of glucose in phosphate-buffer solution. Experimental conditions: electrode area 0.03 cm^2, without a covering membrane, rejuvenated for 15 s at 1200 mV, measured for 15 s at 300 mV in argon atmosphere. Delayed switching of the integrator after 2.5 s. n = number of measuring cycles

Fig. 21 shows a calibration curve which is obtained when the response signal is evaluated and plotted as a function of the glucose concentration in a phosphate buffer solution. The response in this case is the charge which passes through within a measuring time of 15 s at a potential of 1200 mV. The integrator was switched after a delay of 2.5 s, n is the number of measuring cycles using a platinized platinum electrode with an area of 0.03 cm^2. There is a marked concentration dependence and the response signal is reproducible and stable with time. Accordingly, the measurement of glucose concentration appears to be simple and unproblematic. However, difficulties arise in the presence of body substances such as amino acids and urea at their physiological concentrations in phosphate buffer. The measurement of glucose is strongly interfered with by the presence of these substances. Table 4 shows the physiological levels of occurence of amino acids and urea in blood and body fluids. As can be noticed, the physiological levels (low and high) fluctuate considerably.

Table 4. — Minimum and maximum concentrations of amino acids in plasma of adults

Amino acids	Concentrations in mg/dm^3		Difference mg/dm^3 (3 - 2)
	lowest	highest	
1	2	3	4
Alanine	22,2	44,7	22,5
Arginine	8,6	26,3	17,7
Cysteine	4,6	13,5	8,9
Cystine	11,5	33,7	22,2
Glutamic acid	2,5	17,3	14,8
Glycine	10,8	36,6	25,8
Histidine	9,7	14,5	4,8
Isoleucine	4,6	11,5	6,9
Leucine	9,3	17,8	8,5
Lysine	21,1	30,8	9,8
Methionine	2,3	3,9	1,6
Phenylalanine	6,3	19,2	12,9
Serine	6,8	20,3	13,5
Threonine	12,2	29,3	17,1
Tryptophan	5,1	14,9	9,8
Tyrosine	6,5	11,3	4,8
Valine	13,6	26,6	13,0

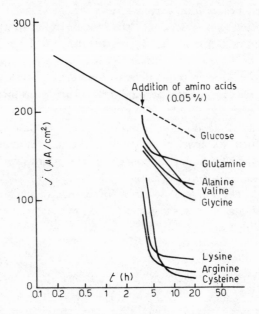

Fig. 22. — Differentiated effect of individual amino acids upon the current density *versus* time curve of glucose oxidation Pt-Ni-foil electrode, phosphate buffer solution pH 7, 22 °C, 0.1 % glucose, 0.05 % amino acid, oxidation potential 0.4 V (r.h.e.)

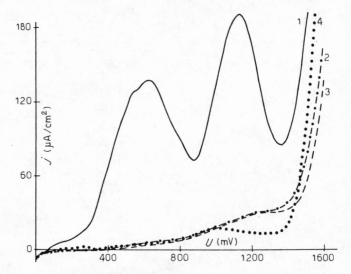

Fig. 23. — Quasi-stationary potentiodynamic current-density *versus* potential curves with a platinized platinum electrode (area 0.28 cm^2), sweep-rate 200 mV/h, argon atmosphere; phosphate buffer solution at pH 7 as the blank electrolyte (BE), 1: BE with 0.1 % glucose; 2: BE with amino acid mixture (AAM) and 0.1 % glucose; 3: BE with AAM and 4: BE alone

Fig. 22 shows the drastic effect of some individual amino acids on the steady-state oxidation of glucose. Two groups can be recognized. The amino acids of the first group depress the glucose oxidation current by 30 %, but the second one by up to 90 %. The figure starts with the current-time behavior of glucose only. The individual amino acids (with a concentration of 0.05 % − half the glucose concentration) were added after three hours (time in logarithmic scale). Thus, the anodic currents associated with glucose are almost completely suppressed when the solution also contains amino acids under physiological concentrations. However, a determinable charge can still be obtained by potential pulsing.

Fig. 23 shows the marked influence of an amino acid-mixture (AAM) also exhibited in slow potentiodynamic curves. Only the anodic part is considered in this quasi-stationary current density *versus* potential curves. The phosphate buffer solution with glucose content (curve 1) shows distinctly two maxima for glucose oxidation. But these maxima disappear completely in the presence of amino acid-mixture under physiological concentrations (curve 2). The presence or absence of glucose made no difference when the elctrolyte contained amino acids (curves 2 and 3). A comparison with the blank electrolyte, BE (curve 4) shows that the oxidation of amino acids begins above 1000 mV.

These results lead to the conclusion that the measurement of glucose concentration should be possible if the access of glucose-amino acid-mixture (AAM) at the electrode is limited by a membrane to such an extent that the arriving amino acids are totally or significantly oxidized. A very active electrode and a relatively impermeable membrane will be required for this purpose. In this connection, the RANEY-platinum catalyst electrode of the type Pt-Ni (1:6), mentioned in the section of biofuel cells, has proved useful with a catalytic layer thickness of 100 μm. Ion-exchanging membranes such as Permion 1025 and membranes of Polysulfone with different degree of sulfonation are well suited as diffusion barrier layers. Polysulfone is especially biocompatible. Table 5 shows a qualitative comparison of diffusion of glucose through several membranes.

The glucose molecule is comparable in its size with that of amino acids. The flow of diffusion is calculated as the diffusion limiting current under the assumption of a complete oxidation of glucose (24 electrons) and a concentration of 0.1 % (column 4).

The diffusion limitation in the case of a relatively slow reaction, such as the oxidation of amino acids, can be achieved by increasing the potential or by increasing the electrode activity, *e.g.* strongly roughening the electrode surface. An improved activity however, would lead in the case of the potential-jump method to an unwanted rise in energy consumption due to the necessary capacitive charge

Table 5. — Diffusion of glucose through membranes

1 Membrane	2 d cm	3 D cm^2 s^{-1}	4 j_{lim}* A cm^{-2}	5 Z_s**
VF 11730				
SARTORIUS	9.1×10^{-3}	1.8×10^{-6}	2.5×10^{-4}	7.7
HYDRON	1.0×10^{-2}	3.3×10^{-8}	4.2×10^{-5}	506
Dialysis tube				
KALLE	4.5×10^{-3}	2.4×10^{-7}	7.0×10^{-4}	14
Cuprophane				
BEMBERG	1.8×10^{-3}	1.8×10^{-7}	1.3×10^{-3}	3
Permion				
(RAI)				
1010 (C)	4.5×10^{-3}	1.1×10^{-6}	3.1×10^{-3}	3
4010 (C)	2.3×10^{-2}	5.8×10^{-7}	3.2×10^{-4}	152
1025 (A)	1.2×10^{-2}	8.7×10^{-8}	9.4×10^{-5}	276
5010 (C)	2.3×10^{-2}	2.3×10^{-8}	1.3×10^{-5}	3841

* $j_{lim} = n \, F \, d \, c_1 / D$ where $c_1 = 0.1 \% = 5.6 \times 10^{-6}$ mol cm^{-3} and $n = 24$

** $Z_s = 0.167 \, d^2 / D$.

exchange. Finally, the transport limitation through a compact less permeable membrane leads at the same time to a longer time constant in detecting the concentration changes of the solution at the electrode surface. The time constant Z_s (column 5) which is also referred as time lag is calculated according to JOST's expression:

$$Z_s = \frac{3 \, d^2 \ln 3}{2 \, \pi^2 \, D} = 0.167 \, \frac{d^2}{D}$$

In considering a controlled dosage of insulin by means of an artificial beta cell, it is essential that the time constant of the sensor should not be longer than that of the glucose level adjustment inside the body (approximately 7 min.).

Thus, if we succeed in achieving a diffusion limitation for the anodic oxidation of amino acids, their influence on glucose oxidation should remain low and uncritical. Although the total concentration of all amino acids amount to half the value of glucose, most of the amino acids are more difficult to oxidize than glucose in the working potential range of 300 - 400 mV.

Fig. 24 shows the concentration variations of amino acids within the physiological range and their influence on the response signal during the glucose concentration measurement, using membranes of two different permeabilities as diffusion barriers. At point 1 in part A, a mixture of amino acids corresponding to the lowest physiological levels of the body fluids was added to glucose-containing solution.

At point 2, the amino acid concentrations were brought up to physiological maximum levels. The marked effect is readily noticeable. An entirely different pattern is obtained by adapting the membrane to the oxidative behavior of the electrode

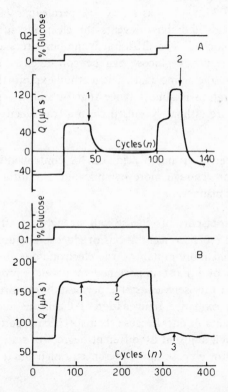

Fig. 24. – Influence of amino acid concentration on the response (Q) of a glucose sensor. n = number of measuring cycles. Electrode area 0.03 cm^2. Rejuvenation at 400 mV for 25 s. Delay time of the integrator 1.5 s. Electrolyte (Tyrode solution) scavenged with a mixture of air (95 %) and CO_2 (5 %); 1. Lowest physiological aminoacid concentration. 2. Highest physiological amino acid concentration. 3. One-half of the highest amino acid concentration.
A) Covering membrane: SARTORIUS 11739; B) Membrane: Permion 1025

in such a manner that all the arriving amino acid molecules are readily oxidized at the electrode as shown in part B. The effect of total amino acids, *i.e.* of a concentration of O raised to physiological maximum values, causes an error of only 13 %. The error is mainly due to the contribution of amino acids to the total current, as indicated by the slight oxidative rise of the curve beyond the point 2. However, when the concentration of amino acids are reduced to half their physiological maximum levels by diluting with glucose solution (point 3), the error is lowered to about 7 %. The concentrations of amino acids fluctuate in practice but not between 0 and the maximum values, and they do not reach the maximum level all at one time. Hence, under normal fluctuation, the error in measurement will probably be around 5 % and will remain below 10 % in any case.

The difference between the calibration curves of glucose and the presence of minimum and maximum amino acid concentrations is likewise small when a membrane, *e.g.* Permion 1025, which has a very low permeability is used. It retards the diffusion of amino acids and thus prevents the electrode surface from being contaminated or easily blocked. The diffusion is inhibited to such an extent that the diffusion limiting current of glucose can be conveniently measured before the electrode is affected by the amino acids. The method exploits the inherent condition that glucose concentration is almost twice that of amino acids and subsequently diffused amino acids are ultimately pulsed clean at the electrode by the process of rejuvenation.

Thus, the influence of amino acids can be controlled in getting the glucose response of the sensor. Though more complicated, the influence of urea can also be tackled in a similar manner.

However, the transport of glucose will strongly be affected leading to low responses because the molar concentrations of glucose and urea are nearly the same and urea is transported more rapidly to the electrode due to its smaller molecular size. Therefore, it is of utmost importance to develop proper membranes with specific properties and precisely adjustable permeabilities in removing the adsorbed layer by subsequent oxidative rejuvenation. Polysulfone membranes, as already mentioned, are quite suitable ones because their specific properties and permeabilities can be defined and well adjusted by different degrees of sulfonation. Fig. 25 shows the calibration curve for glucose in the presence of high and low physiological levels of urea.

By selecting a certain degree of sulfonation (0.5 or 50 %), the polysulfone membrane can be made almost impermeable to glucose to the extent that the reaction at the electrode responds to urea instead of glucose concentration (because of molecular size and diffusion coefficient of urea). In such a manner an urea sensor can be developed and employed to rectify the response of glucose sensor by correc-

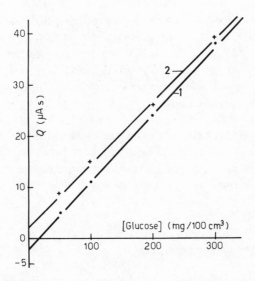

Fig. 25. — Calibration curves for the determination of glucose in Tyrode-solution in the presence of low (curve 1) and high (curve 2) physiological concentrations of urea. Platinum black electrode (0.12 cm^2); Polysulfone membrane; measurement at 400 mV for 50 s and rejuvenation at 1600 mV for 50 s

ting the error in glucose measurement. These aspects are under consideration. The control over these difficulties will enable us to verify the *in vivo* performance of an electrocatalytic glucose sensor.

5.   Transfer of electrical signals and electro-stimulation

5.1.   Preliminary remarks

The problem of interface between the biological tissue or blood and the implant is an essential phenomenon of biocompatibility. The interfacial layer through which an electrical pulse or signal has to be transferred can be depicted as:

electronically conducting	e^- $\longrightarrow$	Biological tissue
implant	$\longleftarrow$ e^-	or blood

Such arrangements find broad application as microelectrodes in recording action potentials at nerve conductors or in the brain and also in trasmitting impulses to nervous tracks or to excitable muscle cells. Potential applications lie in suppressing pain sensations, such as phantom pains, in bladder and in anal incontinence, to stimulate in the event of lameness, in regulating high blood pressure over the sinus carotid nerve, in stimulating dorsal column and in electroacupuncture or electroanesthesia. Above all the pulse transfer to the heart muscle by means of an electronic pacemaker during cardiac disturbances has been widely practiced in medical engineering. Therefore, pacing electrode will be considered as an example of charge transfer inside the body.

The possibility of exciting the muscle by electrical current was demonstrated long ago by GALVANI. Electrical heart stimulation has been known since the last century but it has only become practical since the middle of this century largely owing to the developments in cardiac surgery and in electronics. An implantable pacemaker consists of

a)   a power source,
b)   an electronic circuitry and
c)   a pacing or stimulating electrode.

The first two components (a and b) are cast together in a suitable resin or a metallic encapsulator to form a pacemaker unit (can). The unipolar system, where an electrode is in direct contact with the heart and the second (indifferent) electrode is located somewhere else in the body, is used for internal pacing. A large surface anode made of titanium serves as the indifferent electrode which is usually the can (metallic encapsulator) of the pacemaker. The pacemaker unit (can) is generally implanted in abdominal, shoulder or thorax cavities. The pacing electrode connected to the pacemaker unit by means of an insulated lead is introduced mostly intravenously into the right heart. The pacemaker insertion inside the patient's body is outlined in Fig. 26.

The small current density and low resistance of the large faced anode (Titanium) do not cause muscular twiching or electrochemical corrosion of the metal.

The pulsing voltage should be set in such a manner as to exceed the stimulation threshold; this is the minimum energy required to release an effective contraction of the heart. The stimulation threshold increases from the time of implantation, passes through a maximum during the course of a few weeks and reaches a constant value. There is no direct relationship between the initial and later threshold value. A rise of the threshold above the available energy of the pacemaker is called *exit block*.

Clinically implantable pacemakers are operated at a mean current of 30 μA with two batteries (Li-I$_2$-cells) each having an approximate voltage of 2.7 V. This pacemaker voltage of 5 V employed today is nearly 100 times that of the potential change which is required to excite a nerve or muscular fibre (*ca.* 60 mV). The energy of 10^{-5} Ws which is practically needed is a thousand times the minimum energy

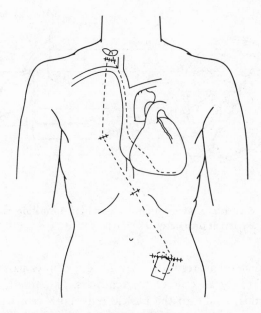

Fig. 26. — Implantation of an electronic cardiac pacemaker and the electrode
lead-wire through *vena jugularis* into the right atrium

at which the heart is stimulated. Thus, the electrical power applied determines
the life period of the pacemaker and the pacemaker size is essentially determined
by its power source. The size of the power source can considerably be reduced
(thus, miniaturizing the pacemaker) and yet a longer service life maintained, if the
energy for the stimulating pulse is utilized economically (optimization of power
consumption).

5.2.   *Cardiac uses*

5.2.1. Cardiac stimulation. – In a sound heart the sinus node takes charge of the
natural pacemaker function. It is anatomically located in the right atrium close to
the mouth of *vena cava* superior as shown in Fig. 27.

The impulse which initiates the heart-beat starts in the sinoatrial (S-A) node,
which is less a point than a region. From the S-A node an electrical excitation wave
spreads through the muscle of the atria to the atrioventricular (A-V) node, which
lies at the junction of the atrium and ventricles. The excitation wave causes the
atria to contract. At the same time the impulse is conducted through the A-V node
to the bundle of HIS and its branches to the ventricles. The bundle of HIS is a

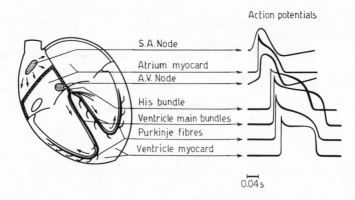

Fig. 27. — Schematic representation of excitation formation and propagation systems and the relevant time-delayed action potentials

conducting tract that originates in the A-V node; its fibres pass into the ventricles and then excite virtually all cells of the ventricles. As a result, after electrical impulses pass along this bundle to the muscle cells, both ventricles contract almost simultaneously. By contrast, in the atria the electrical wave passes progressively from one muscle cell to the next at a finite, relatively slow rate. As a result, a gradual coordinated contraction wave passes over the atria. It is precisely this delay in propagating the excitation which enables the atria to empty the blood into the ventricles. A long refractory period of approximately 300 ms in the myocard prevents the self feed-back of the muscles by means of circulatory excitation. When the sinus node fails, the A-V node can take charge of the pacemaker function with 60 pulses/min or even the bundle of HIS and PURKINJE fibres with 25 to 45 pulses/min. But these lower intrinsic rates are insufficient for a normal cardiac output. Hence, the artificial pacemaker is used to provide a more rapid rate.

In the actual cardiac pacing, energy can be saved by two means:

i)   by minimizing the electrochemical losses during the passage of current, *i.e.* non-polarizable electrodes, and

ii)   by eliminating the post-operative threshold rise; the threshold rises because of the growth of non-excitable fibrotic tissue around the electrode. Therefore, a tissue-compatible electrode material is required to overcome this problem. The pacer electrodes most commonly used today are made of noble metals such as Pt and especially Pt-Ir.

5.2.2. Energy saving electrode for cardiac pacing. — Fig. 28 shows the voltage behavior of some electrode materials at a galvanostatic square-wave current impulse of 1 ms duration measured in an aqueous solution of 0.15 M NaCl at 22 °C. A slow voltage rise depending on the material used is observed after a jump from 0

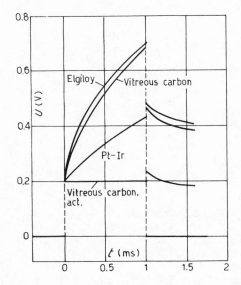

Fig. 28. — Potentials of different electrode materials during galvanostatic stimulating pulse as a function of time, measured with a current density of $j = 10$ mA cm^{-2} in a solution with 0.15 M NaCl at 22 ° C

to nearly 0.2 V. This rise occurs as a consequence of different double-layer capacitances which are also dependent on the frequency and the potential. The rise is high with Elgiloy and unactivated vitreous carbon, lower with Pt-Ir and no longer detectable with activated vitreous carbon as electrodes. The threshold energy values of different electrodes evaluated from impedance measurements are schematically represented in Fig. 29.

In this representation, if we assume a mean threshold current of 0.15 mA with a semi-spherical electrode of 1.2 mm radius and a resistance of 720 Ω or 573 Ω, a value of 2.1×10^{-8} Ws or 1.6×10^{-8} Ws is obtained as a threshold energy for a non-polarizable electrode. The lowest individual values lie at 3×10^{-9} to 5×10^{-9} Ws. The energy consumption is particularly low for electrodes prepared from TaC and activated vitreous carbon. These results of *in vitro* measurements were encouraging to test the *in vivo* performance of activated vitreous carbon as a pacing electrode.

In Fig. 30, the *in vivo* threshold current behavior of different electrodes is recorded. The electrode tips mostly semispherical in shape, having a radius of 1.2 mm, were placed in pairs 10 to 15 mm apart in gluteal muscle of cats. The electrodes were contacted with an Elgiloy coil and the lead was covered with a silicone tube. The threshold was measured galvanostatically with square-wave current impulses of 1 ms duration. In order to prevent an electrochemical charge transfer

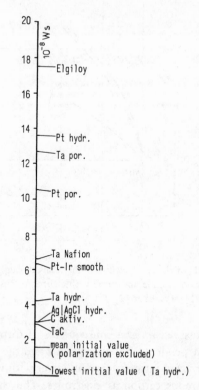

Fig. 29. — Threshold energy values calculated from impedance measurements with smooth, porous, Hydron and Nafion coated electrodes

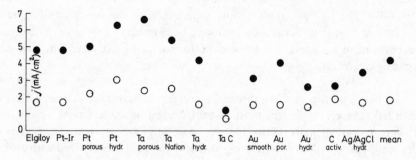

Fig. 30. — Threshold currents initially (○) and after 4 weeks of implantation (●)

at the electrode, a current was passed in the opposite direction during the pulse interval so that the charge was compensated. A threshold analyzer was used as the square-wave generator. The stimulating current was measured oscillographically with the help of a resistance of 10 or 100 Ω. The amplitude (current) at which the periodic muscular contractions provoked by stimulation first appear was considered to be the threshold. The voltage behavior at the threshold pulse was also observed. The square-wave shaped current pattern enables us to differentiate between spreading resistance at the stimulating electrode and the electrode polarization, and to deduce quite separately the specific resistance at the electrode surrounding.

In this manner, the threshold measurement with all electrodes were made at implantation and at explantation (after 4 weeks of implantation). The threshold at the explantation is, in general, considerably higher than at the beginning. Actually, it is not the low threshold change but the low chronic threshold which is decisive. Low values of chronic thresholds have been found with Pt-Ir, smooth Au, TaC and activated vitreous carbon. The mean values of initial threshold voltages varied between 65 mV and 328 mV while the thresholds at explantation were generally higher, ranging between 117 mV and 445 mV. The values, for instance, for Pt-Ir and activated vitreous carbon were 196 mV to 265 mV and 157 to 200 mV respectively (TaC: 65 mV and 117 mV).

The histological examination of the electrode beds and the muscle surrounding the electrode tips also revealed the excellent tissue compatibility of vitreous carbon. The mean thickness of connective tissue layer formed around Pt-Ir tip was 325 μm (ranging between 125 to 800 μm). On the other hand, connective tissue layers of only 25 to 50 μm in thickness were formed around carbon tips.

This satisfactory performance in skeletal muscle was also confirmed in further animal tests. The finished stimulating electrodes made of activated vitreous carbon tip were inserted into the canine heart through the *vena jugularis*. Threshold measurements using galvanostatic current impulses at implantation and explantation were carried out in a similar manner as described in skeletal muscle experiments. In addition, the stimulation response and its sequence was checked by recording the electrocardiogram. The electrodes were connected to a variopacemaker which was also implanted in the neck region of the animal (dog). During the implantation period, the variopacemaker remained normally inhibited at a heart frequency above 70 beats/min. But in recording the threshold voltage, it could be converted to fixed rate, however, by applying a magnet, thus increasing its frequency to 100 beats/min and reducing its normal output voltage of 5, 2.5 or 1.25 V, in a series of 15 equal steps. The 0-impulse is detectable in the electrocardiogram and the number of impulses can be counted from this point. The threshold was characterized by the lowest vario-impulse generating contractions of the heart.

Fig. 31 shows the *in vivo* performance of carbon tip electrodes in different dogs. The initial threshold values of < 175 mV and < 350 mV respectively increase but remain mostly between 350 mV and 525 mV. The results of Pt-Ir electrodes are shown in Fig. 32. The electrode in dog 7 was dislocated but the other two electrodes adhered well to the tissue. The threshold value increased in dog 6 to a value between 875 mV and 1050 mV and in dog 10 between 1400 mV and 1575 mV. These results indicate that at least 40 - 50 % of the energy can be saved by using activated vitreous carbon instead of the conventional Pt-Ir electrodes.

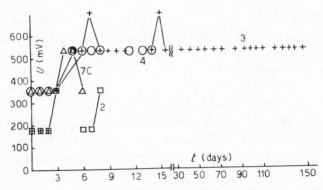

Fig. 31. — Threshold behaviour of well-placed carbon-tip electrodes in four different dogs (variopacemaker step: 175 mV)

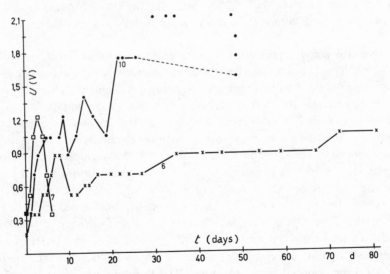

Fig. 32. — Threshold behavior of Pt-Ir electrodes in three different dogs (variopacemaker step: 175 mV)

Figs. 33, 34 and 35 show the histological sections of the electrode placement and tissue reactions of carbon tip and Pt tip electrodes in cardiac muscle. A connective tissue layer of less than 100 μm in thickness separates the apparently intact muscular layer from the carbon electrode (Fig. 34). On the other hand, strong irretations in the surroundings of Pt-Ir electrodes were observed (Fig. 35). Even the muscular tissue far away from the electrode is affected and this may explain the considerable rise in threshold. The threshold rise cannot be exclusively explained by the growth of connective tissue and its thickness surrounding the electrode. The formation and the thickness of the fibrotic tissue may contribute to a 30 % rise in the threshold value. The ohmic resistance of the fibrotic capsule is smaller than that of the intact muscular tissue. Hence, changes in muscular tissue leading to inactivation must also take place at a greater distance from the electrode.

Clinical tests with carbon tip electrodes implanted in 373 patients showed distinctly a better performance over Pt tip electrodes; yielding improved results up to 50 %. Thus, for instance, the mean threshold voltage over 18 months was 0.85 V for carbon tips. The mean value for Pt-tips over the same period was above 1.6 V. Thus, the electrochemical considerations and their consequent application have led to a significant and practical contribution to the clinical cardiac pacing.

6.   Electrochemical system in artificial kidney

The artificial kidney for the treatment of patients with end-stage chronic uremia relies on hemodialysis or hemofiltration of various toxines present in blood, such as urea, creatinine and uric acid, into a buffered salt solution known as dialysate or filtrate. The removal of these molecules, *e.g.* urea, and, at the same time, maintaining the electrolyte balance of the dialysate or filtrate is of critical importance in an artificial kidney treatment. One possible method of removal is based on electrochemical degradation or decomposition of the organic toxine into non-toxic products.

If conditions in an electrochemical cell are properly chosen, urea can be oxidized completely into harmless products such as nitrogen gas, carbon dioxide and hydrogen according to the simplified scheme:

$$6\,Cl^- \longrightarrow 3\,Cl_2 + 6\,e^- \qquad \text{anodic reaction}$$

$$(NH_2)_2CO + 3\,Cl_2 + H_2O \longrightarrow N_2 + CO_2 + 6\,H^+ + 6\,e^- \qquad \text{anolyte}$$

$$6\,H^+ + 6\,e^- \longrightarrow 3\,H_2 \qquad \text{cathodic reaction}$$

$$\overline{(NH_2)_2\,CO + H_2O \longrightarrow N_2 + CO_2 + 3\,H_2 \qquad \text{overall reaction}}$$

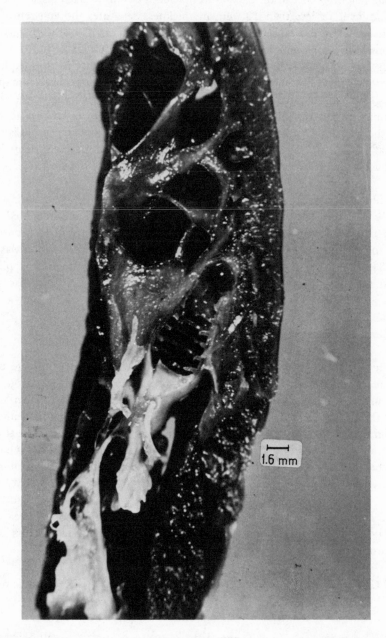

Fig. 33. — Electrode bed together with the outline of the fixation rings in the cardiac muscle (f = 5)

Fig. 34. − Electrode bed of a well placed carbon tip electrode in dog 3 (f = 40). Minimal induction of connective tissue after implantation

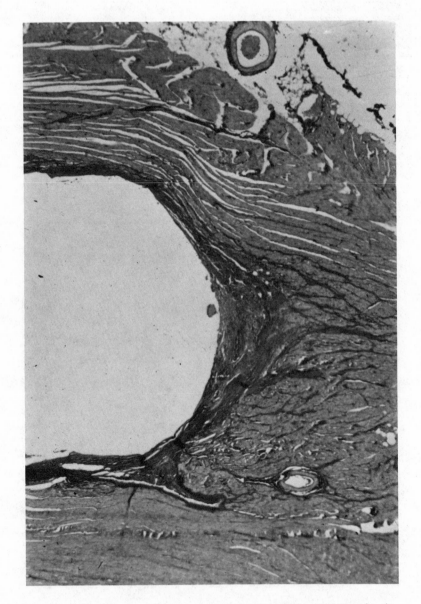

Fig. 35. — Tissue reaction of a Pt-Ir electrode in dog 10 (f = 40). Granular tissue protruding into the surrounding muscle

Electrical variables such as current (current density) and voltage, and also the electrode activity as well as the rate of flow of the dialysate or filtrate will determine the course and efficiency of such an electrochemical overall reaction. As indicated by the partial reactions, it is the generated chlorine which oxidizes urea in the anolyte.

A schematic principle for indirect electrochemical oxidation of urea is outlined in Fig. 36. The electrolyte containing urea and chloride is streamed into the anode compartment, where elemental Cl_2 is formed at the electrode by electrochemical oxidation of Cl^- ions. The elemental chlorine reacts with urea, thus, leading to the formation of CO_2 and N_2. Since a cation-exchanging membrane is used to separate the electrode compartments, Na^+ ions migrate from the anolyte to cathode compartment. The pH-value of the streaming electrolyte drops during the passage through the anode compartment. The solution reaches the cathode compartment where H_2 is formed at the electrode. The pH value of the catholyte increases and the electrolyte leaves the electrochemical cell finally as a neutral deureated solution. In the actual practice, the ideal behavior of anode is restricted because oxygen, along with chlorine, will also be liberated at the electrode. The gases Cl_2 and O_2 transported into the cathode chamber can be reduced at the cathode. However, the formation of hypochlorite, which is toxic, must be prevented or at least its traces removed from the electrolyte.

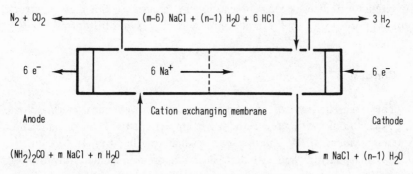

Fig. 36. — Principle of indirect electrochemical oxidation of urea (according to RICHTER, MUND and WEIDLICH)

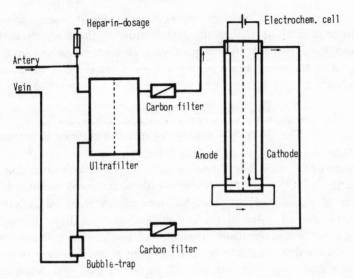

Fig. 37. — Hemofiltration and regeneration of the filtrate by indirect electro-chemical $(NH_2)_2$ CO-oxidation (according to RICHTER, MUND and WEIDLICH)

The flow diagram in Fig. 37 shows the scheme of hemofiltration and regenera-tion of the filtrate by means of indirect (extracorporeal) electrochemical oxidation of urea. At a constant flow rate of the filtrate, the concentration of urea decreases with increasing current density of the electrode. With a proper design of the cell and with a combination of a few such electrochemical cells, urea can be cleared or removed at a rate of 7.2 g/h. The deureated filtrate can then be brought back into body circulation after taking precautionary measures. Such an electrochemical system has proved very useful in preliminary trials of hemofiltration.

7.   Conclusion

The research, state of art and scope of electrochemical methodologies have been described only with a few elaborate examples such as I) biogalvanic and biofuel cells (energy sources), II) glucose sensor (diabetes therapy), III) stimulating elec-trodes (cardiac pacing) and IV) detoxification of urea in hemofiltration (artificial kidney). There are still a number of other possibilities of exploiting electrochemical principles and properties in clinical and biomedical practice. All these aspects cannot be considered here. However, mention must at least be made of some methods which are of practical importance.

1. Nerve stimulation for pain relief, treatment of hypertension and angina, for hearing aid etc., in short for applications in neural prostheses, and also brain stimulation as an aid to the diagnosis of brain pathology and as a prosthetic device, are most actual topics where different kinds of microelectrodes are used which are characterized by electrochemical phenomena. Also electro-anesthesia and electronarcosis are based on electro-stimulative effects.

2. Implantable electro-osmotic driving pump for controlled and safe release and delivery of drugs.

3. In setting electrochemical parameters, especially the criteria, for biomaterials (blood and tissue compatibility).

All these aspects are under intensive pursuit all over the world, demonstrating the practical importance of electrochemistry and its contribution in preserving and improving the quality of life. In conclusion, some references are cited in the following to enable interested colleagues to gain not only a general information but also a more detailed and deeper insight into all those aspects that were described and mentioned in this lecture.

References

Books

G. MILAZZO, *Elektrochemie I: Grundlagen und Anwendungen,* Birkhäuser Verlag, Basel, Boston, Stuttgart, (1980).
C.D. FERRIS, *Introduction to Bioelectrodes,* Plenum Press, New York, London, (1977).
H.A. MILLER and D.C. HARISON (Editors), *Biomedical Electrode Technology: Theory and Practice,* Academic Press, New York, London, (1974).
J. KORYTA, *Methods for Electroanalysis in vivo in Electroanalytical Chemistry,* A.J. BARD (Editor). M. Dekker, New York, Basel (1966), Vol. 11, p. 85.
P.W. CHEUNG, D.G. FLEMING, M.R. NEUMAN and W.H. KO (Editors), *Theory, Design and Biomedical Applications of Solid State Chemical Sensors,* CRC Press, Florida, USA, 1978.
K.D. HEPP, W. KERNER and E.F. PFEIFFER (Guest Editors), *Feedback-controlled and Preprogrammed Insulin Infusion in Diabetes Mellitus,* Hormone and Metabolic Research, Supplement Series No. 8, Georg Thieme Verlag, Stuttgart and New York, 1979.
L.L. HENCH and E.C. ETHRIDGE, *Biomaterials — The Interfacial Problems,* in *Advances in Biomedical Engineering,* J.H.U. BROWN and J.F. DICKSON (Editors),

Academic Press, New York, London, (1975) p. 35.

F.T. HAMBRECHT and J.B. RESWICK (Editors), *Functional Electrical Stimulation: Applications in Neural Prostheses,* in *Biomedical Engineering and Instrumentation,* Marcel Dekker, Inc., New York, Basel, (1977), Vol. 3.

H.J.TH. THALEN and C. MEERE (Editors), *Fundamentals of Cardiac Pacing,* Martinus Nijhoff Publishers, The Hague, Boston, London, (1979).

W. IRNICH, *Elektrotherapie des Herzens − physiologische und biotechnische Aspekte,* Fachverlag Schiele & Schön, Berlin, (1976).

T.M.S. CHANG (Editor), *Artificial Kidney, Artificial Liver and Artificial Cells,* Plenum Press, New York, London, (1978).

A. LIMOGE, *An Introduction to Electroanesthesia,* University Park Press, Baltimore, London, Tokyo, (1975).

F. von STURM, *Implantable Electrodes,* in *Topics in Bioelectrochemistry and Bioenergetics,* G. Milazzo (Editor), J. Wiley, Chichester, New York (1979) p. 191.

Articles

M.I. BABB and A.M. DYMOND, Electrode Implantation in the Human Body, *Bull. Prosth. Res.* **10**, 51 (1975).

E. WEIDLICH *et. al,* Animal Experiments with Biogalvanic and Biofuel Cells, *Biomat. Med. Dev. Art. Org.* **4** (3&4), 277 (1976).

J.R. RAO *et. al,* Electrochemical Behavior of Amino Acids and their Influence on the Anodic Oxidation of Glucose in Neutral Media, *Biomet. Med. Dev. Art. Org.* **6** (2), 127 (1978).

U. GEBHARDT *et. al,* Development of an Implantable Electrocatalytic Glucose Sensor, *Bioelectrochem. Bioenerg.* **5**, 607 (1978).

E. WEIDLICH *et. al,* Threshold Measurements Using Stimulating Electrodes of Different Materials in the Skeletal Muscles of Cats, *Med. Progr. Technol.* **7**, 11 (1980).

G.J. RICHTER *et. al,* Chronic Threshold of Stimulating Electrodes: Comparison of Activated Vitreous Carbon with Conventional Platinum-Iridium Electrodes in Animal Tests, *Med. Progr. Technol.* **8**, 67 (1981).

K. MUND *et. al,* Development of a Non-Polarizable Stimulating Electrode for Implantable Cardiac Pacemakers, *Siemens Forsch.- u. Entwickl.-Ber.* **8**, Nr. 4, 227 (1979).

M. FELS, Recycle of Dialysate from the Artificial Kidney by Electrochemical Degradation of Waste Metabolites: Small-Scale Laboratory Investigations, *Med. Biol. Eng. Comput.* **16**, 25 (1978).

G. LUFT *et. al,* Electro-osmotic Valve for the Controlled Administration of Drugs, *Med. Biol. Eng. Comput.* **16**, 45 (1978).

P.N. SAWYER, Application of Electrochemical Techniques to the Solution of

Problems in Medicine, *J. Electrochem. Soc.; Rev. News,* **125**, (10), 419C (1978).

P. BAUERSCHMIDT and M. SCHALDACH, The Electrochemical Aspects of the Thrombogenicity of a Material, *J. Bioeng.* **1** (4), 261 (1977).

CONCLUDING REMARKS

GIULIO MILAZZO

Institute of Chemistry of the Engineering Faculty
of the University, Via del Castro Laurenziano 7 — 00161 Rome, Italy

This course is the first one on bioelectrochemistry attended by professors from five, perhaps six nations if we consider Dr. Rao coming from India, and participants from seven countries. Attendance was about 50% by people having a biological background and 50% by people with a physico-chemical background.

We thank all the lecturers for the subjects illustrated with utmost clarity and logical development. But a very special word of thanks is due to Professor Metzner who was submitted to a marathon of six hours and was able to speak without interruption, without consulting a single piece of paper, and with such a logical chain of arguments that I wonder if such a feat is really possible for a human being. Thank you very much Professor Metzner.

We listed the subjects included in the program, all in the field of redox reactions, in a sequence which had (at least we tried to give it) a certain sense of logical development: thermodynamics, energetic balance, capture of solar energy as the first source of life, kinetics, particularly important reactions in living bodies (enzymes, mediators, photobiology, damage from radiation), sophisticated new techniques, and research on particularly important points like the respiratory chain concluding with some biomedical applications. This means that the search for truth in scientific research when applied with good faith always brings some advantage to mankind. The biomedical applications are a really good example of what I mean.

Why this marathon of four, five, six hours for some speakers? I owe you an explanation. This course was programmed originally for ten days and, for reasons independent of the will of the organizers, it had to be reduced to five days of real working time. Therefore, the speakers were compelled to condense what they had to say in a very short time. Perhaps this has led to some inconveniences in the sense that very highly condensed concepts were not immediately captured by some of those not familiar with the subject matter. This problem, however, will be eliminated by the publication of all the lectures in perhaps a more extended form, and in any case with an extensive list of references so that any doubts about the concepts presented can be solved and the way to better understanding of particular points will be made available.

Is it possible to enumerate some of the results of this course? I think so. First of all, the attendance was fifty-fifty percent between biologists and physico-chemists, including among the biologists people specialized in zoology, botany, physiology or other branches of life sciences, and among physico-chemists those specialized in physics, biochemistry, physical chemistry, and so on. This is already one of the results of our efforts to bring biologists and physico-chemists together into closer cooperation. It is really impossible to remain enclosed in an island or in a *turris eburnea*, as our Latin forefathers used to say, and not consider developments in sister branches to solve the problems presented by the phenomena occurring in living bodies. This is already a good result in itself. We hope that in the future this interpenetration of scientists with a biological extraction and those from to so-called exact sciences will be more and more a concrete result. I am sure that each and every one of those attending this course has learned something here.

There is another aspect of utmost satisfaction. Perhaps you notice that from time to time I counted those present. I was pleased to note that average attendance was usually over 90%. This is a very high percentage. I have never seen such high continual attendance in specialized courses lasting even six or seven hours of lectures per day. I must congratulate you on your persistance.

Another additional positive result. I think this course was appreciated by those attending it and we had some very explicit and clear requests on future editions. Therefore we have given some initial thought to this possibility and in the Council of the Bioelectrochemical Society we are considering another course to be centered on some electrochemical aspects of biological membranes.

This course will perhaps be conducted in November 1984. There are two reasons for this apparently late time point. First al all, the Centre is booked solid until this time. The second reason why we are considering 1984 is that this course is at the same time one of the activities and main events of the Bioelectrochemical Society. The Bioelectrochemical Society holds its general meeting every two years. In fact, we had our general meeting in the summer of this year. The next one is scheduled to take place in 1983. It is not advisable to hold the general meeting and the course in the same year. Travelling costs today are very high and people who wish to attend both events, the general meeting and the course, would find that very costly. Therefore, we will try to rotate every other year in the future and I hope for a long time: every second year the course and the general meeting in the alternate years.

In this sense perhaps I could give you some further information on the Bio-electrochemical Society which is responsible for this course as well. When opening this course I did give you some information but perhaps it would be a good idea to repeat what I said. During my career as lecturer of Electrochemistry at the Rome University and at the same time in the Chemistry Laboratories of the (Italian) Institute of Health, after working in this Institute for about 40 years, I became aware of the fact that many and perhaps the majority of the phenomena in living bodies are electrochemical in nature or in origin. And then in 1971, in an attempt to pursue this idea I organized the first bioelectrochemical symposium, which at that time went under the heading *Biological Aspects of Electrochemistry*. It was very favourably received in the sense that when it was over, the participants, like what is happening here today, asked where we would hold the second symposium. Our French colleagues were immediately ready to offer French hospitality for the second one. At the end of that symposium the scenario was the same. The third meeting was held in Germany, the fourth in the United States, the fifth in East Germany, the sixth in Israel, and the seventh will take place in 1983. We are already working on the eighth Congress to take place most likely in Japan. You can see that interest is expanding everywhere in the world.

This, however, is not the only activity of the Bioelectrochemical Society. We soon found it advisable to have a periodical, a review which could serve as a forum where biologists, physicochemists, biophysicists, and so on, could discuss and diffuse the results of their research. In the libraries of biological institutes, a whole group of physico-chemically oriented periodicals like *Journal of the Electro-chemical Society, Zeitschrift für Physikalische Chemie, Berichte der Bunsen-Gesell-schaft, Journal de Chimie Physique, Transactions of the Faraday Society* etc. are not to be found. Conversely in the libraries of physico-chemical or physical institutes we don't find *Biochimica and Biophysica Acta, Biochemical and Biophysical Research Communications* and many other biologically oriented periodicals. That was the reason for launching a periodical which could be present in both libraries and improve the mutual knowledge of biological problems viewed from a physico-chemical or physical point of view and from a biological point of view, as well. In effect, the new periodical *Bioelectrochemistry and Bioenergetics* is printed today in about 1 500 copies, and I deem it essential to quantify the diffusion of these period-icals. Actual circulation is about 1 200 copies. This is a good figure and illustrates the increase of interest in this branch of scientific research.

A further stage was the organization of this course which we hope will become a regular event of the Centre and the Bioelectrochemical Society. The publication of the lectures delivered here could also be considered, from another point of view, as the starting point for a *Treatise on Bioelectrochemistry*. A very important consideration in the sound development of any new branch of scientific research is the need for a Society, a School, regular Meetings and a Treatise. We have already achieved the first three, and in this way we hope to start with the fourth element.

There is also a further point to be considered in a positive fashion. People attending the Erice school found so much to be of mutual interest that they are thinking of promoting closer cooperation between physicists, bio-chemists, and biologists. I consider this to be a very important point because problems of living matter are so complex and so difficult that it is impossible to investigate them from only one point of view, with only one kind of technique, and with only one type of technical and theoretical knowledge on hand. The physicist must talk with the biologist, the physiologist must discuss things with the electrochemist, and so on. This intention to pursue closer cooperation in research is another good result.

I have one last point to mention and it is an important result not only of this course but of all the courses held in this Centre. It is the mutual understanding among people coming from different parts of the world. Unfortunately in our world there are still active prejudices which represent a barrier to better understanding between people. Particularly, young people, who are more open than people of my age, are open to this mutual knowledge and will contribute in a substantial way to better understanding. For this reason I thank you for coming and I hope you will come again, and in a larger number.

LIST OF PARTICIPANTS

Directors of the course

Prof. Giulio MILAZZO
President of the Bioelectrochemical Society
Piazza G. Verdi 9, 00198 Rome, Italy

Prof. Martin BLANK
Dept. of Physiology, Columbia University
630 W 168 Street, New York, N.Y. 10032 USA

Speakers

Prof. Hermann BERG
Institute of Microbiology
Academy of Sciences of the German Democratic Republic
11 Beutenbergstr., 69 Jena, German Democratic Republic

Prof. Martin BLANK

Prof. René BUVET
Laboratory of Biochemical and Electrochemical Energetics
University Paris Val de Marne
Ave. Général De Gaulle, Créteil, 94010 France

Prof. Bruno Andrea MELANDRI
Istituto e Orto Botanico Università
Via Irnerio 42, 40126 Bologna, Italy

Prof. Helmut METZNER
Institute of Chemical Plant Physiology of the University
Corrensstr. 41, 7400 Tübingen, German Federal Republic

Prof. Giulio MILAZZO

Prof. H. Wolfgang NÜRNBERG
       Institute of Chemistry IV Applied Physical Chemistry
       Nuclear Research Center
       517 Jülich, German Federal Republic

Dr.   I. Raghavendra RAO
       Research Laboratory Siemens
       582 Erlangen, German Federal Republic

Prof. David R. WILSON
       Dept. of Biochemistry and Biophysics
       School of Medicine G 3
       Philadelphia, Penn. 19104 USA

Participants

Angela AGOSTIANO
Institute of Physical Chemistry
Via Amendola 173, 70126 Bari, Italy

Adriana BOLASCO
Chair of Physical Chemistry, Insitute of Pharmaceutical Chemistry
Città Universitaria, 00199 Roma, Italy

Claudio BOTRE'
Chair of Physical Chemistry, Institute of Pharmaceutical Chemistry
Città Universitaria, 00100 Roma, Italy

Andrea CEGLIE
Institute of Physical Chemistry
Via Amendola 173, 70126 Bari, Italy

Maurizio CIGNITTI
Istituto Superiore di Sanità
Viale Regina Elena 299, 00100 Roma, Italy

Barbara CZOCHRALSKA
Dept. Biophysics, Inst. Experimental Physics, University of Warsaw
93 Zwirki Wigury, 02085 Warsaw, Poland

Mauro DEGLI ESPOSTI
Istituto ed Orto Botanico Università
Via Irnerio 42, 40126 Bologna, Italy

Klaus FRISCHKORN
Institute of Applied Physical Chemistry, Nuclear Research Center
517 Jülich, German Federal Republic

Olle INGANÄS
Dept. Physics and Measurement Technology, Linköping University
58183 Linköping, Sweden

Grits KAMP
Dept. Microbiology, Groningen Biological Center
Kerlan 30, 9751 NN Haren, Netherlands

Douglas KELL
Dept. Botany and Microbiology, University College of Wales
Aberystwyth, SY23 3DA, UK

Egon KOGLIN
Institute of Applied Physical Chemistry, Nuclear Research Center
517 Jülich, German Federal Republic

Lucia Raffaella LANGUINO
Istituto Chimica Biologica
Via Amendola 165/A, 70100 Bari, Italy

Halvor LERVIK
Parliamentary Assembly of the Council of Europe
Strasbourg, France

Iris LOEW
Max Planck Institute of Biophysics
Kennedy Allee 70, 6000 Frankfurt, German Federal Republic

Adriana MEMOLI
Chair of Physical Chemistry, Institute of Pharmaceutical Chemistry
Città Universitaria, 00100 Roma, Italy

Israel MILLER
Membrane Research Dept., Weizmann Institute of Science
Rehovot, Israel

Eberhard NEUMANN
Institute of Biochemistry, Max Planck Society
Martinsried, München, German Federal Republic

Daniela PIETROBON
Institute of General Patology
Via Loredan 16, 35100 Padova, Italy

Günter PILWAT
Arbeitsgruppe Membranforschung, Institute of Medizin,
Nuclear Research Center
517 Jülich, German Federal Republic

Massimo PIZZICHINI
Laboratory of Physical Chemistry, Centro Casaccia - CNEN
Via Anguillarese km 1.2, 00123 Roma, Italy

Enrico POLASTRO
Soc. Solway, Laboratoire Central
Rue de Rausberk 310, 1120 Brusselles, Belgium

Roberto SANTUCCI
Center of Molecular Biology CNR, Institute of Biological Chemistry
Città Universitaria, 00100 Roma, Italy

Craig SCHENCK
Dept. Biologie, Service Biophysique, Centre Saclay
91191 Gif-sur-Yvette Cédex, France

Jean Marie SEQUARIS
Institute of Applied Physical Chemistry, Nuclear Research Center
517 Jülich, German Federal Republic

Patrick SETA
Laboratory of Physical Chemistry CNRS
Boite Postale 5051, 34033 Montpellier, France

Joan SMITH SONNENBORN
Dept. Zoology and Physiology, University Wyoming, P.O. Box 3166
Laramie, Wyoming 82071, USA

Mario ZORATTI
Institute of General Patology
Via Loredan 16, 35100 Padova, Italy

SUBJECT INDEX